• Drew MacArthur Elicker • Port Townsend • Summer • 1996

—THE PARTICLE CONNECTION—

The Most Exciting Scientific Chase
Since DNA and the Double Helix

CHRISTINE SUTTON

Simon and Schuster   New York

To Carlo, who made this book possible,
and to Terry, who makes all things possible

Published by Simon and Schuster
A Division of Simon & Schuster, Inc.
Simon & Schuster Building
Rockefeller Center
1230 Avenue of the Americas
New York, New York 10020
SIMON AND SCHUSTER and colophon are registered trademarks of
Simon & Schuster, Inc.
Manufactured in the United States of America
10   9   8   7   6   5   4   3   2   1
Library of Congress Cataloging in Publication Data

Sutton, Christine.
    The particle connection.
    Bibliography: p.
    Includes index.
    1. Particles (Nuclear physics)   I. Title.
QC793.2.S88   1984   539.7'2   84-10595

ISBN: 0-671-49659-X

Contents

1

The Fruits of Discovery

Anthropologists label modern man as *Homo sapiens*, or 'wise man'. The first members of this species lived over 35,000 years ago, and they were no doubt wise in the sense of being able to apply their experience and knowledge in a judicious manner. But was it wisdom alone, encouraged by an advantageous environment, that led to the species' dramatic development over the ensuing years? Wisdom in itself does not necessarily lead anywhere; it often merely produces the best means of dealing with a set of circumstances within a known framework. Perhaps what made *Homo sapiens* so strikingly different was his curiosity; perhaps the anthropologists should call us 'curious man'.

Even 30,000 years ago our ancestors must have been curious. Wherever they lived they must surely have wondered about fire, about thunder and lightning, about rain. They would have noticed the differences in materials, between rocks and water, between the soft hair on an animal's back and the coarse bark of a tree. They must have asked themselves why night follows day, what makes the sun shine and how the world began.

The search for answers

The mythologies and ancient religions of all societies are riddled with attempts to answer these questions. Stories of creation abound; gods are associated with inexplicable and often terrifying natural phenomena, such as volcanoes and solar eclipses. Natives of the Hawaiian archipelago saw the unpredictable eruptions of the active volcanoes on their islands as bursts of temper from the goddess Pele, who liked to live in the fire-lined pits. The ancient Egyptians believed in the sky goddess Nut, who swallowed the sun at sunset, but gave birth to it again at dawn.

The gods, humans and beasts that populate the myths were the only means 'curious man' had to explain the world about him; they reflect his ignorance, his lack of more appropriate descriptive means. But as *Homo sapiens* developed, so he discovered better techniques to explain the workings of our universe. The curiosity that could at first only ask questions but not answer them, except through the agency of imaginary creatures, led ultimately to algebra, geometry and the concepts of systematic measurement and experiment; in other words, the tools of scientific discovery. It is the same curiosity about the world around us that drives the scientists of today – from the anthropologists who are pasting together a picture of those very people from whom our curiosity descends, to the physicists who seek the answer to age-old questions about the nature of the universe.

The old name for physics – natural philosophy – probably gives a better sense of the work in which many modern physicists find themselves engaged. Questions regarding the nature of matter, the meaning of reality, the origins of space and time in the universe seem often more appropriate to the popular image of a philosopher than to that of a physicist making measurements in a detached, objective way. But physicists in laboratories throughout the world are contemplating and seeking to find the answers to some of the most basic questions we can ask. Why does the sun shine? What are the basic building blocks of matter? How did the universe begin and how will it end?

One of the most awesome achievements of science in the twentieth century may well prove to be that physicists began to find answers to some of these questions, for our understanding of the fundamental nature of matter is in a better state than it has ever been. Moreover, the latest results from experiments that probe the constituents of matter show that not only is this understanding all-embracing in terms of the phenomena it seeks to explain – like the Hawaiian story of the goddess Pele – but it seems as though it may be right.

This remarkable lack of modesty is based on some impressive work by many physicists which has taken place particularly since the early 1930s. It is true that such confidence is dangerous, and that there have been other times when scientists believed they had arrived at the 'complete' theory of the physical world. However, the latest theory has not only a neatness and a wholeness that make it particularly attractive, but also a remarkable predictive power. And it is the experimental proof of these predictions that makes the physicists so sure of themselves.

The language and the mathematical detail of the new physics would no doubt mystify natural philosophers from times gone by, but the basic concepts and framework upon which the theories are constructed would surely meet with their approval. Any explanation of the world about us must begin by addressing certain fundamental questions, and these remain the same as they were centuries ago. One of the most basic questions, and one that must enter the minds of most of us when we are children, is to ask what the world is made of.

The building bricks of matter

A common school of ancient Greek philosophy, perpetuated well into the Middle Ages, pictured the world as containing four 'elements': earth, fire, air and water. A fifth pure, perfect element – the quintessence – was reserved for the heavenly bodies alone. In some ways the philosophers who embraced this view showed still more conceit than the physicists of today. They did not stop at explaining the physical world in terms of their four elements, but went on to characterize human behaviour according to four related 'humours' or fluids which determined the disposition of the mind and body. Thus were physics, physiology and psychology all united in one grand world view. Those philosophers, obsessed with the number four, may well take heart from the modern view, that four forces are required to give the universe the properties we observe. But more of this later.

A more pedantic look at the question 'What are we made of?' came from a particular school of thought active in Greece around the fourth and fifth centuries BC, when the philosopher Leucippus and his student Democritus developed the theory of atomism. They proposed that at some level matter comprises identical indivisible particles, which they referred to as atoms, from the Greek word meaning, literally, something that cannot be cut.

The concepts of atoms and elements, which originated with the ancient Greeks, came together in the birth of modern chemistry in the late eighteenth and early nineteenth centuries. John Dalton, the weaver's son from Cumbria, arrived at his atomic theory through considering the ways in which different elements combine. The old philosophy based on four elements – or even only two elements, sulphur and mercury, as proposed by the alchemists of the fifteenth century – had given way slowly to the idea of elements as simple, unmixed substances. In Dalton's *A New System of Chemical Philosophy*, published in 1808, he listed twenty elements, which he supposed were all composed of atoms, or 'extremely small particles' (Figure 1). The 'ultimate particles' of each element were all identical; in particular they were all of the same weight, which varied from one element to another. From this basis Dalton was able to explain the laws for the combination of elements into compounds.

Atoms, as introduced by Dalton, remained the indivisible ultimate constituents of matter, just as Leucippus and Democritus had imagined them, for the best part of a century. But in 1897, J. J. Thomson,

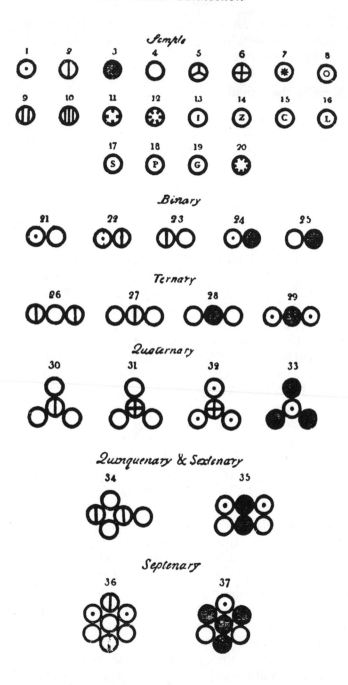

Figure 1   *A plate from John Dalton's* A New System of Chemical Philosophy *(1808) shows his symbols for the different atoms that form the elements, and the way they combine to form simple molecules (courtesy Vivien Fifield)*

Cavendish Professor in the University of Cambridge, wrote in the *Philosophical Magazine* (vol. 44) that he had discovered 'matter in a new state, a state in which the subdivision of matter is carried very much further than in the ordinary gaseous state: a state in which all matter . . . is of one and the same kind; this matter being the substance from which all the chemical elements are built up.' Thomson had found the electron, a minute negatively charged fragment of matter, that weighs only 9.1×10^{-31} kg,* and which we know exists in the atoms of all the elements. We also know, as Thomson rightly believed, that the atoms of the elements are different aggregations of identical, more basic constituents, although it was to take another thirty years or so to determine the exact nature of those constituents. And though the building blocks of the elements – alike for each element, but differing from one material to the next – proved to be patently divisible, the name atom stuck, and new names were found for the more fundamental constituents.

If the atom contains negatively charged constituents – electrons– then so must it contain an equal amount of positive charge, to make atoms, and matter, neutral overall. Thomson pictured the tiny electrons as negative 'raisins' distributed through a positively charged 'pudding'. But in 1911, Ernest Rutherford, working at Manchester University, made a different proposal. Experiments in which alpha particles from a radioactive source were directed at gold foil showed that the positively charged alphas often passed straight through the foil and were only rarely deflected through large angles. Rutherford proposed that the positive charge of an atom is concentrated at its centre, while the electrons travel around it in distant orbits. Thus, according to Rutherford, most of the atom is empty space, and alpha particles can generally pass right through gold foil; occasionally, however, they come close enough to the nucleus to be deflected through large angles, just as is observed.

So, in the first decades of the twentieth century, the structure of the atom began to become clear. Each element, be it a metal, like copper, or a gas, like chlorine, consists of atoms whose weight characterizes the element; thus far Dalton was right. What he never knew was that each atom comprises negatively charged electrons and an equal number of positively charged particles – called protons – to make matter electrically neutral overall. The number of protons (and electrons) increases one at a time to give the elements increasing atomic weight. Thus hydrogen, the lightest element, has one electron and one proton; helium, the next lightest element, has two electrons and two protons; and so on to uranium, the heaviest naturally occurring element, which has ninety-two electrons and ninety-two protons.

* See Appendix I for an explanation of this notation for representing very small and very large numbers.

The protons are about 1800 times heavier than the electrons and reside at the heart of the atom, in the nucleus, which is typically 10^{-13} cm across. The electrons orbit the central nucleus, attracted by the opposite charge of the protons, rather like a miniature solar system, though this is an analogy that should not be taken too far.

But this picture is not quite complete, as nuclear physicists discovered in 1932. The nucleus also contains neutral particles – neutrons – which have a similar mass to protons. They help to keep the nucleus bound together, partly by diluting the positive charge of the protons. Being like-charged, the protons would otherwise tend to fly apart, due to the very electrostatic force that keeps the electrons within the atom.

Neutrons differ from protons and electrons in another way: they are unstable. A 'free' neutron, away from the confines of the forces at play in the atomic nucleus, lives on average for only fifteen minutes or so, after which it transmutes into a proton, producing an electron at the same time. The neutral neutron thus gives birth to a positive proton and a negative electron, and so conserves overall electric charge. This process underlies the radioactivity of many unstable nuclear species, in which a neutron can transmute even though bound within the atom, for in changing the balance of neutrons to protons it takes the nucleus on a route to a more stable configuration. Radioactivity is simply the process of nuclei trying to find stability.

The decay of the neutron gives rise to one further particle, in addition to the proton and the electron. This particle carries with it the balance of momentum and energy, thus ensuring that these quantities, like electric charge, remain the same before and after the reaction. It is electrically neutral but has little or no mass at all. The particle is known as the neutrino and was suggested theoretically by Wolfgang Pauli in the early 1930s, more than twenty years before it was eventually proved to exist in 1956.

Thus in the 1930s physicists knew of four particles: the proton, the neutron, the electron and the neutrino. The mystical number four seemed to have some say in the running of the universe after all! But even as this simple pattern emerged, there were hints that nature was not quite so straightforward.

In 1928 Paul Dirac, a theoretical physicist at Cambridge University, found two sets of solutions to equations describing the possible energy states of electrons. One set of solutions corresponded to electrons observed in the real world. The second set of solutions was more enigmatic and seemed to correspond to electrons with negative energy! Dirac proposed that nature did in fact observe both sets of solutions to the energy equations, and that the extra set in effect described an as yet unobserved particle with the electron's mass, but with opposite electric charge. This positively charged particle was named the positron and was discovered in 1932.

Although Dirac's work in 1928 was concerned with electrons, its conclusions could be extended to protons, and therefore implied the existence of a negatively charged version of the proton. Indeed, it became apparent that for all matter in the real world – corresponding to one set of solutions – there should exist an 'opposite' form of matter, or 'antimatter'. Thus the electron has a corresponding antielectron – the positron; and the proton should have a corresponding antiparticle, the antiproton. So while the proton, neutron, electron and neutrino were sufficient to explain the real world about us, it seemed that nature's underlying structure contained more particles, which were normally hidden from view.

Experiments in the 1950s were to confirm the existence of the antiproton, but in the meantime the nuclear physicists started to uncover a new 'world within a world'. In studying the interactions of cosmic rays – high-energy protons and nuclei from outer space – with atomic nuclei in the upper atmosphere, physicists began to discover new types of particle. Like the neutron, these particles proved to be unstable, although their lifetimes were much shorter, being a few hundred-millionths of a second (10^{-8} s) or even less. The new particles decayed eventually to protons, electrons and neutrinos, but they lived long enough to show that they had properties similar to the known particles. So one could identify the new unstable particles as being related in some way to the more familiar stable particles.

The general exploratory work with cosmic rays was backed up and improved upon by more specific studies in laboratories. With the aid of particle accelerators, experimenters could take protons up to high energies with velocities close to the speed of light, and so produce collisions mimicking those of cosmic rays in the earth's atmosphere. The laboratory work had the added advantage that the accelerators could produce far more high-energy particles than could easily be studied in cosmic-ray experiments, and so it was possible to learn rapidly many details of the behaviour of the new 'elementary' particles. By the 1960s it had become clear that there were literally hundreds of unstable particles. Chaos seemed to be taking over from the orderly picture of four fundamental particles.

The restoration of order was only just round the corner. It came in the 1970s with further experiments which showed that the proton, neutron and most of the new particles were – like the atoms of the chemical elements – not truly indivisible, but composite. The search for the atoms of the ancient Greeks had moved to a new level.

The new level of structure reveals that the proton, neutron and their unstable relatives are built from the same basic units: the quarks. To manufacture the huge variety of subatomic particles found in experiments, it seems that there are probably six types of quark. These combine in different ways to form the observed particles: the proton and neutron each consist of different combinations of three quarks, selected in different ways from only two of the quark types. If we say, for argument's sake, that the proton contains two quarks of type 1 and one of type 2, then the neutron contains two of type 2 and one of type 1. Other particles exist, albeit not for long, which contain three quarks all of type 1, or three all of type 2, and so on.

The quarks seem to possess one quality that differentiates them from those objects that have previously laid claim to being the true elementary particle. While experiments can knock atoms out of a material, and protons and neutrons out of the nuclei of atoms, no experiment has succeeded in driving a single quark out of a proton. Furthermore, there seem to be good theoretical reasons why this should be so, and this leads many physicists to wonder whether we are in fact at the level of the ultimate constituents of matter.

How do the electron and the neutrino fit into the quark 'model'? They are not, it seems, made from quarks; instead they appear, as far we can tell, to be indivisible. Therefore they really do deserve the tag of 'elementary' particle. Moreover, among the new particles discovered in cosmic-ray experiments and at accelerators, there are a few that behave very much like the electron and the neutrino. These particles include the muon, which although unstable forms one component of the cosmic rays that reach the ground – it is the product of decays of more exotic objects formed in the initial collisions. There is also the tau (τ), produced in experiments for the first time in 1975. The tau and the muon both resemble the electron, so much so in fact that they seem like heavier 'photocopies' of the electron. There are also two more types of neutrino, other than the kind produced in neutron decay. Thus in total there seem to be six particles that are not built from quarks. These are collectively named the leptons, from the Greek for 'small' or 'slender'.

Six quarks, six leptons: the magic number four seems to have given way to six. These particles, together with their corresponding antiparticles, are sufficient to build the whole physical universe – as far as we can tell. Perhaps, nearly 2500 years after Leucippus and Democritus considered the question of the composition of the world, we have at last found the answer (Figure 2).

Or have we? Even if we can lay claim to having discovered the ultimate constituents of matter, the indivisible atoms, we can hardly say that we have a complete understanding of the construction of the universe. Given a bag full of quarks and leptons, could we in principle build a universe?

The answer, of course, is no, for our universe kit would be incomplete.

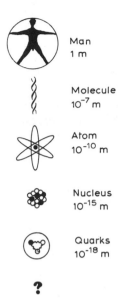

Man
1 m

Molecule
10^{-7} m

Atom
10^{-10} m

Nucleus
10^{-15} m

Quarks
10^{-18} m

?

Figure 2 *As we look more and more closely at the structure of matter, we perceive new layers, each built from smaller constituents than the last. Man is formed from complex molecules comprising many atoms. The atoms contain electrons orbiting a nucleus, itself built from protons and neutrons. These in turn appear to consist of quarks; are they the true building blocks of matter?*

How, for instance, could we 'glue' the quarks and leptons together? How would we form them into different objects? The world of our everyday experience does not appear as a three-dimensional mosaic of identical tiny pieces. It consists of a huge variety of substances, from the ground beneath our feet to the air we breathe; from the simple crystals of a snowflake to the complex structure of the human body. What holds these seemingly different things together? And on a larger scale, what holds our planet in orbit around the sun? What keeps immense clusters of similar suns together in galaxies, visible across the vastness of space?

The universal glue

We cannot in general measure atoms and molecules without the aid of sophisticated techniques; the moon, the sun and some of the planets, on the other hand, are visible to the naked eye and have been observed for thousands of years. So, it is hardly surprising that the large-scale structure of our local region of the visible universe was the first to surrender to a sound physical theory. The inspiration for the theory came, so the story goes, to a young man walking in his mother's orchard in the heart of Lincolnshire. In 1666, the twenty-four-year-old Isaac Newton was passing time at home while the university in Cambridge was closed during the years of the Great Plague. He was later to claim that it was while watching apples fall from the trees that he began to ask himself whether the effects of gravity, which made the fruit tumble to the ground, would extend out as far as the moon's orbit. Was gravity responsible for keeping the moon in its orbit about the earth?

Whatever the truth about Newton's inspiration, there can be no doubt about the size of his eventual achievement, summarized in his law of universal gravitation. This says that there is but one force that produces a mutual gravitational attraction between all bodies, whatever size and however far apart. The strength of the force between two objects depends only on the product of their masses divided by the square of the distance between them, all multiplied by a constant, symbolized as G, which, as far as Newton was concerned, had the same value in all parts of the universe. Whereas the values of the masses and the distance between the objects set the relative strength of the gravitational force, the value of G, which can be measured experimentally, sets its absolute strength.

With Newton's law of gravitation we can work out the force that keeps the earth orbiting the sun, that makes apples fall from trees, and that holds billions of stars together in the galaxies unheard of in Newton's day. Newton's audacity lay in extrapolating gravity on earth not only out to the moon but beyond; in conceiving of a universal force that would attract all bodies whatever their nature. Such audacity was surely the hallmark of genius. With Newton gravity became something not just peculiar to the earth, or even to the planets and other members of the solar system, but a force acting between all manner of objects.

Newton did not know of atoms as we think of them today, but he would have appreciated the popular image of a solar system in miniature, the electrons orbiting the central nucleus. The tiny planetary system is a useful analogy, but the differences between Newton's celestial mechanics and the behaviour of the atom are huge. The electrons orbit the nucleus at relatively far greater distances than the planets circulate around the sun; the force at work in keeping the atom together is clearly not gravity, although it bears some likenesses to gravity.

An understanding of the force that holds the atom together came fifty years or so before the internal structure of the atom began to come to light. And as with Newton's achievement with universal gravitation, the theoretical insight came only after good observational groundwork existed.

While certain aspects of electricity and magnetism had been apparent for centuries, it was only in the latter part of the eighteenth century that physicists such as Charles-Augustin Coulomb and Alessandro Volta began to perform precise experiments on electricity and magnetism, and to lay the foundations of what has become one of the best understood branches of physics. In the following decades Hans Oersted, André Marie Ampère and Michael Faraday were to discover and explore the interweaving of electric and magnetic phenomena: how an electric current produces a magnetic field, and how a current is generated in a wire moving through a magnetic field. But the genius who put all this work in order, who found the theoretical nutshell into which he could pack a host of physical phenomena, was the Scottish physicist who later became professor at Cambridge, James Clerk Maxwell.

Maxwell published his *Dynamical Theory of the Electromagnetic Field* in 1865, two centuries after Newton had walked through his mother's orchard and pondered his 'system of the world'. In this work, Maxwell provided a single, mathematically based theory for the interlinked phenomena of electricity and magnetism – or electromagnetism. The theory requires not one fundamental equation, like Newton's did, but four. But in the way that Newton achieved more than his initial aims, so did Maxwell. Newton set out to extrapolate the earth's gravity to the moon's orbit; he finished with a universal law applicable to all kinds of bodies. Maxwell's bonus was that his theory not only encompassed all electromagnetic behaviour but that it also included all optical phenomena. Light, it seems, had to be a form of electromagnetic wave propagating through space according to the dictates of Maxwell's equations. His theory requires such waves, and requires their velocities to be the same as that measured for light.

Just as Newton's universal law describes a force of gravitational attraction between all objects, so Maxwell's equations give the electromagnetic force between all electrically charged bodies, moving and stationary. Indeed, one of Maxwell's equations embodies the electrostatic law of force between two charges, which Coulomb had written down in 1785. This law states that 'the repulsive force between two small spheres charged with the same sort of electricity is in the inverse ratio of the square of the distance between them'.

Coulomb's law bears strong resemblances to Newton's universal law of gravitation, but there is also one striking difference. Electric charge can take one of two opposite values, which we define as positive and negative. If the charges are both of the same sign, then Coulomb's force is positive; but if the charges have opposite signs, then the force is negative. This in turn reflects the fact that for charges of the same sign the force is repulsive – 'like charges repel'; for charges of opposite sign, however, the force is attractive, as in the case of gravity, and 'unlike charges attract'.

Electrons, as J. J. Thomson found, are electrically negative; the nucleus is positive. Thus it is Coulomb's law, rather than Newton's law of gravity, that binds the atom together. Maxwell's electromagnetism, albeit modified to deal with the peculiar surroundings of the subatomic environment, rules the behaviour of the charged atomic solar system. Electricity, long before it was tamed by man, was essential to his existence in order to build up the rich variety of atoms and complex molecules that form the building blocks of life.

So far, so good. Gravity holds the earth on its yearly path around the sun, and keeps our feet placed firmly on the ground; electromagnetism binds the electrically charged constituents of the atom. But when we come to the atomic nucleus we reach a new difficulty. Experiments show that the nucleus seems not to be a single positively charged conglomerate of matter, but a number of positively charged units. The charge on each of these units is equal and opposite to that of the electron, and with equal numbers of electrons and positive units the atom is electrically neutral. The positive unit seems to be the nucleus of the simplest atom, hydrogen; we call it the proton. The problem comes when we ask why the nucleus does not fly apart. After all, like charges repel, according to Coulomb's law. What we seem to need is a third kind of force; and in the way that electromagnetism is stronger than gravity, then so must this new nuclear force be stronger than the electrostatic repulsion between the protons. Indeed, physicists have named it simply the strong nuclear force.

As with the other forces, a good theoretical explanation of the strong force depends on good experimental observations. Only in the past few years, over half a century since experiments began to indicate the composite nature of the atomic nucleus, are theorists settling on a prescription to put the strong force on a sound footing. Unlike the cases of gravity and electromagnetism, there has been no single brilliant development by one man, rather the collective effort of a number of brilliant minds, working on a force that in many ways is quite unlike the others.

One reason why it has taken a relatively long time to come to terms with the strong force is that the nucleons – the protons and their neutral companions in the nucleus, the neutrons – are not the source of the strong force. The strong force originates instead in the quarks, the constituents of the nucleons. The force binding the protons and neutrons in the nucleus is but a residue of the force between the quarks – the strong force that holds the quarks together within the protons and neutrons that make up the nucleus. This echoes the electrostatic binding of atoms within certain types of molecule. In this case, the force holding the molecule together is a residue of the Coulomb force that keeps the electrons orbiting the atomic nuclei. And in the same way that these molecules form from electrically neutral atoms, so the nucleus forms from nucleons that are neutral in terms of the strong force.

The strong force appears to originate in a new type of charge, which is called colour because, like the primary colours of light, it seems to come in three varieties. The quarks carry colour charges and are bound together within the nucleons by the strong force; the colours of the quarks in each nucleon cancel out so that the nucleons themselves are neutral.

The analogy between the colour forces at work in the nucleus and the electrostatic forces at play in a molecule works to a certain extent, but it breaks down when we come to consider an equation describing the strong force. This force appears not to decrease inversely with the square of the distance, as in Newton's law of gravitation, and Coulomb's law of electrostatic force. On the contrary, it seems to increase, being relatively weak when the quarks are close together, but becoming stronger as they move apart. We do not know the nature of the strong force well enough to write down precise equations, but it seems that at small distances – small relative to the size of a proton! – it resembles the Coulomb force. However, at larger distances it must change so as to increase sufficiently to keep the quarks tied together within the particle they form.

The force with a difference

It seems, then, that there are three forces holding matter together, from the strong force operating in the confines of the atomic nucleus, through the electromagnetic force holding atoms and molecules together, to gravity which operates the celestial clockwork.

Surprisingly perhaps, this is not the complete picture. There is another force. However, this manifests itself not so much in holding matter together but in allowing it to disintegrate and transmute from one form to another. This force, like the strong force, operates only over dimensions smaller than the atomic nucleus. Fortunately, it is many, many times weaker than the strong force, otherwise there would be no nuclei and no matter to form us or the universe we inhabit. But the fact that the weak force does allow certain species of nuclei to decay is vital to our existence, for it forms crucial links in the chains of nuclear reactions that fuel the sun and build up heavy elements from the simplest nuclei.

The weak force is responsible for the decay of the neutron into a proton, electron and neutrino; this was the reaction that Enrico Fermi studied in the 1930s when he developed the first theory of the force. Later studies showed that the same force underlies the decay of other particles, for example, the muon, which decays into an electron and two neutrinos. Just as Newton found that gravity was a universal force at work in a wide variety of circumstances, physicists in the late 1940s and early 1950s discovered that the weak force operates on a wide variety of particles. Indeed, it is the only force apart from gravity that the neutrinos feel, for as they have no electric charge, they are not influenced by the electromagnetic force; nor do they feel the strong force, which operates only on quarks.

Twentieth-century research into the nature of matter has thus revealed

GRAVITY operates across the universe. The same force that makes apples fall off trees, binds thousands upon millions of stars together in galaxies like our own Milky Way. It is weakest of the four forces

The WEAK nuclear force operates only at distances of 10^{-15} m or less, but plays a crucial role in the reactions that fuel the Sun and other stars. It underlies the radio-activity of many unstable nuclei

The ELECTROMAGNETIC force binds negative electrons to a positive nucleus in in the atoms that form all matter. It under-lies all manner of phenomena from the radiation of sunlight to earth, to radio and TV

The STRONG nuclear force binds the elementary quarks together within the protons and neutrons of the atomic nucleus. It is unusual in that it becomes stronger at larger distances, thus confining the quarks within the larger particles

Figure 3   *The four forces known in nature, ranging in strength from the feeblest, gravity, to the strong nuclear force. Each force plays a crucial role in forming our universe*

two more forces, over and above gravity and electromagnetism, which are necessary to bind matter together, and to bring about the nuclear reactions that build up the variety of elements in the universe. The magic number four seems relevant after all: four forces are needed to provide nature's diversity, to build atoms from quarks and leptons, to make grains of sand and complex protein molecules from atoms, and to form the stars and galaxies (Figure 3). But not only does nature display an immense diversity; the four forces themselves are quite different. Gravity holds galaxies together but is far too feeble to hold an atom together. The weak force makes many forms of matter unstable, and defines to a large extent what may and may not be, yet its influence extends no farther than a hundredth the diameter of a proton, and it is far weaker than electromagnetism, or the strong force. Why are the forces so different? And why, for that matter, are quarks so different from leptons?

The quest for unification

There is a feeling among physicists that by searching for common features in a diverse world, they can learn more about nature's underlying principles. The basis for this rationale is sound enough. In seeking a common thread between falling objects here on earth and the orbit of the moon, Newton set off down the road that led him to a universal law of gravitation. In tying down the links between electricity and magnetism, Maxwell developed a coherent theory not only of these phenomena but also of light.

The early twentieth century saw the rise of another genius who sought simplicity through unification in physics: Albert Einstein. In his theory of relativity Einstein united space and time, adding time to the three dimensions of space to create the four dimensions of space–time, the fabric of the universe on which matter is embroidered. His general theory of relativity takes Newtonian gravity and develops its real meaning in the context of space–time. Whereas to Newton time and space were absolute, Einstein showed that space and time, like beauty, lie in the eye of the beholder. Time at the top of a skyscraper runs ever so slightly slower than at the bottom, due to the difference in gravity; a particle travelling close to the speed of light has a distorted view of its stationary surroundings. Most of Einstein's work on relativity for which he is so well known was done by the time he was in his thirties. Much of the rest of his life was spent in trying to search for another, equally important, unification. Having put gravity on a sound theoretical basis, valid in all corners of the universe, he turned to trying to incorporate general relativity with Maxwell's electromagnetism.

Einstein did not succeed, and physicists today are still a long way from a single theory that describes the effects of gravity and of electricity and magnetism. However, success has come from a slightly different direction. Attempts to improve upon Fermi's work have uncovered a theory for the weak force that is both physically plausible and mathematically sound and which naturally incorporates electromagnetism. Despite their manifest superficial differences, the weak and the electromagnetic forces appear to be but facets of a single force.

This bald statement contains a tremendously profound concept. A single theory unites phenomena as disparate as the radioactive decay of a neutron and the generation of a television picture; as far apart as the burning of the sun and the flickering of a compass needle. This amazing achievement is sufficient to spur the particle physicists to further audacity. Why not incorporate, they ask, the strong force? Certainly the unified 'electroweak' theory has suggested the way forward in dealing theoretically with the strong force, so that it is better understood than ever before in the past fifty years. The idea has been to devise a theory of the strong force that is mathematically similar to the electroweak theory. The result is the theory that regards colour as the source of the strong force. The next step is to take the colour theory of the strong force and unite it with the electroweak theory. This leads to the creation of a 'grand unified theory' which provides explanations for three of natures forces for the price of one theory.

The ultimate achievement will, of course, be to succeed where Einstein failed: to include gravity in the 'totally unified theories'. It is the heady thought that they are on the brink of this very success that drives many of today's physicists and has set the whole field from particle physics to cosmology ablaze with excitement. Why are the physicists so confident? Why so now, much more than before?

The ideas of grand unification and total unification rest firmly on the success of the electroweak theory. While indirect evidence for this theory grew during the 1970s, it was only in 1983 that direct evidence first appeared in experiments. The electroweak theory predicts the existence of new particles, which are unlike the quarks and leptons, but which are in effect the 'light' of the weak force. Whereas the electromagnetic force gives rise to the radiation of radiowaves, X-rays, visible light, and so on, the weak force in a sense radiates the new particles.

The observation of these new particles vindicates the electroweak theory. The story of these particles and the way they were discovered unfolds in the following chapters. This is the story of the W particle, its rarer companion the Z^0 particle, and the prospect of a complete understanding of the universe.

2

A Look Inside

The philosophers of ancient Greece were generally content simply to think about the fundamental nature of matter. Their minds and their imaginations were their laboratories; what was unthinkable was not possible. By contrast, the natural philosophers of the past 300 years or so – the physicists – have had an altogether different attitude. They have investigated the world to find out about the way it is, and have used their imaginations to work out how it comes to be that way. Experiment is the exploratory tool that has revealed the motion of the planets, the interconnection of electricity and magnetism, and the nature of the atom. In each and every case experimental data have provided the scaffold upon which the edifice of theory is built.

How do we investigate the nature of matter? The ancient Greeks imagined being able to cut matter into smaller and smaller pieces until indivisible units – the atoms – are eventually found. Such an experiment, although simple in concept, turns out to be precluded in practice by the extremely small size of the ultimate components; they are far, far smaller than anything we can cut in any conventional sense. However, dissecting an object systematically is not the only way to discover its constituent parts.

Imagine finding a vase full of unknown articles, but with a neck too small to extract the contents simply by inserting a hand. One way to find out what the vase contains might be to drop it so that it smashes open on the floor. An alternative might be to throw stones, or something similar, at the vase in the hope of breaking it open; a catapult would probably be necessary to put sufficient force behind the projectile. Another equally destructive method of analysing a substance comes into play when a chemist conducts a 'flame test', burning a powder to discover its nature by the colour of the flame: copper gives a green tinge, while sodium burns bright yellow.

These techniques are not too dissimilar, for with both the vase and the powder we use some form of energy to unlock the secrets of the object's composition. We would use gravitational energy in dropping the vase to break it open, or elastic energy generated from a catapult in hurling a projectile at the vase. The flame test requires thermal energy to initiate the burning process.

These examples are both analogous to methods used to probe to the very heart of matter, deep within the atom. We can simply harness brute force to try to knock things out of the atom, and see what we find in the resulting debris. Or we can be a shade more subtle and look for indirect effects, like the colouring of the flame by light emitted from atoms of a particular element.

A lesson from spectroscopy

The flame test is a crude form of spectroscopy – the analysis of matter via the study of the radiation it emits when energized. Ordinary sodium atoms emit no light; we see the material by the light it reflects. But once energized – or 'excited' – the atoms emit a characteristic yellow light, colouring the flame when common salt (sodium chloride) burns, or providing the yellow glow of a sodium street lamp, in which the atoms gain energy from an electric discharge. The basic spectroscopic technique is to observe the spectrum of light an excited body emits, as when a prism disperses 'white' light from the sun into the broad range of colours from red through yellow and green to violet. The yellow light from a sodium lamp is also dispersed by a prism, not into a broad band of varying colour, however, but into two bright lines. The general yellow colour of the flame from common salt exposes the presence of the element sodium in the burning compound; but what do the lines in the spectrum from a sodium lamp reveal about the structure of sodium itself?

Light, as Maxwell discovered, is a form of electromagnetic wave, which travels through free space at a velocity of 300 million metres per second (3×10^8 m/s). The wavelength of the light is the distance between successive peaks in the wave motion, and the colour varies with the wavelength. Red light has the longest wavelength of visible light, about 0.7 millionths of a metre (700 nanometres), while violet light has the shortest with a wavelength of around 0.4 millionths of a metre (400 nm). So, when a prism disperses light, it is spreading it out according to its wavelength, bending the red component the least and the violet component the most. The lines in the sodium spectrum show that the element emits light at only certain specific wavelengths. A similar study of the light from a mercury lamp, or from a hydrogen discharge, or indeed from any element, will show a similar spectrum of lines, each peculiar to the element. The spectrum in effect

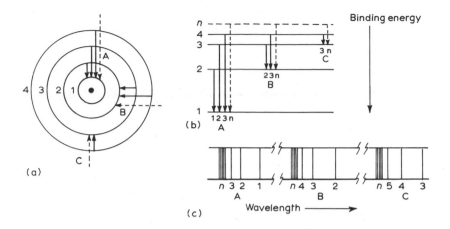

Figure 4 *Electrons radiate energy when moving from outer orbits in an atom to inner ones (a). Each orbit has an associated energy, so moves between different orbits can be depicted as jumps between different energy levels (b). Jumps from outer orbits to each inner orbit give rise to a series of lines in the spectrum (A, B, C . . .), where shorter wavelengths correspond to larger energy differences (c)*

provides a 'fingerprint', and the pattern of lines contains clues about the nature of the element.

We now know that the different wavelengths for a particular element are emitted as electrons jump from outer orbits in the atoms to ones closer to the nucleus (Figure 4). In exciting the atoms we give them energy which the electrons can use to move to more energetic orbits, farther from the nucleus. But once in the 'excited state' the electronic configuration is unstable, and the electrons must lose their newly gained excess energy by emitting radiation and returning to their normal orbits. The wavelengths corresponding to the different lines in the spectrum are characteristic of jumps between different pairs of orbits: a larger jump sheds more energy, which is radiated as light at a shorter wavelength. In this way, the spectral fingerprint identifies the element in terms of its detailed internal structure, as the wavelengths reveal the precise energies of the particular arrangement of electrons.

This book is not really concerned with atomic physics, but this very brief digression about atomic spectroscopy serves to illustrate the way in which a systematic study of some phenomenon that is little understood can lead to clarification on a grand scale. Although we now understand the spectra of the elements in terms of electron orbitals, when scientists such as the Swiss mathematician Johann Balmer began systematic studies of atomic spectra in the 1880s, little was understood about the nature of the atom. But the details of these observations and the pattern that emerged were the basic groundwork on which the Danish physicist Niels Bohr was later, in 1913, to build his theory of the atom. Bohr's model of an atom with one electron provided the first quantitative explanation for the spectrum for the simplest element, hydrogen. It gave correctly the wavelengths of the lines observed in terms of transitions between the specific orbits available to the single electron. And the essence of Bohr's model remains at the heart of our theoretical understanding of atomic spectra to this day.

This diversion into atomic spectroscopy also serves to illustrate another point: how energizing an object – in this case, an atom, or rather a host of atoms – can lead to new discoveries about the contents of that object – in this example, the energies of the atomic electrons. A similar process of discovery by the exploitation of energy and the subsequent systematization of observation has led to a profound understanding of the structure of particles like the proton, and the theory of quarks and the strong interaction.

The tools necessary to explore the proton are far more complex than those used in the early part of the twentieth century to reveal the structure of the atom. Electrical discharges have been replaced by huge particle accelerators; prisms and photographic film by complex particle detectors. The energizing part of the experiments is generally akin to throwing stones at a vase; the detailed study of the contents often resembles atomic spectroscopy.

The subatomic catapult

The accelerator is perhaps the single most important item in a modern particle-physics experiment. This is the 'front end', supplying the energy that is used to reveal the fundamental constituents of matter; this is the catapult that flings the stones at the nuclear vase. The projectiles most commonly used are electrons and protons. As the constituents of atoms, they are easy to supply, and being electrically charged, they are relatively easy to accelerate.

A charged particle in an electric field experiences a force which accelerates the particle, so that it gains energy. In a field due to another stationary electric charge, the particle feels a force given by Coulomb's law, and depending on whether the charges are like or unlike, the particle will move away from or towards the source of the field. In modern accelerators the electric field is supplied by an electromagnetic wave, which can be regarded as an oscillating electric field perpendicular to a similarly oscillating magnetic field (Figure 5). Exactly how the electromagnetic wave transfers

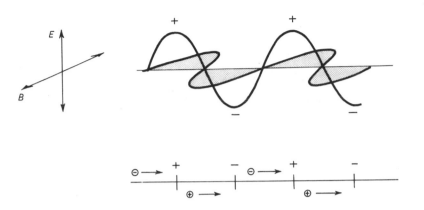

Figure 5   *An electromagnetic wave consists of mutually perpendicular oscillating electric (E) and magnetic (B) fields. A charged particle will be accelerated by the electric-field component of the wave only during the appropriate part of the cycle. Positive particles are accelerated towards more negative regions of field, negative particles towards more positive regions; in the opposite parts of the cycle, the particles are decelerated if not shielded somehow from the electromagnetic wave*

energy to the charged particle depends on the particular type of accelerator, but the principle remains the same. The direction of the electric-field component of the electromagnetic wave must be such that it gives the particle an added boost in the direction in which it is already travelling. So a positive particle, such as a proton, must see a field that is more negative in front of it so that it is attracted and accelerated forwards, towards the negative region. Since electromagnetic fields oscillate between positive and negative, the trick with an accelerator is to make certain that the particle encounters the right part of the wave motion and is accelerated.

A major concern with any particle accelerator is to attain as high an energy as possible. A practical way to do this, rather than becoming involved with dealing with very high electric fields, is to make repeated use of a more modest field. In a *linear accelerator* the particles are directed through successive regions with the same strength of electric field; in the machines called *synchrotrons* they are made to follow essentially circular paths so they pass repeatedly through the same regions of electric field. In both cases the outcome is the same: the particles reach high energies through the net effect of many small 'pushes'.

The fact that the particles can be made to follow curved paths is another consequence of the electromagnetic behaviour of matter. Whereas a stationary electric charge feels no force in a magnetic field, a moving charge does feel a force, in a direction perpendicular both to the direction of the magnetic field and to the direction of travel. Thus a charged particle travel-

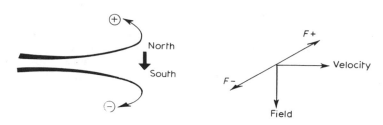

Figure 6   *An electrically charged particle feels a force when moving through a magnetic field. The direction of this force, F, is perpendicular to the field and to the motion of the particle, and depends on the charge as shown. The net effect is to force the particle to follow a curved path*

ling between the poles of a magnet will be deflected, the way it is deflected depending on the sign of its charge (Figure 6).

Today's largest accelerators are synchrotrons, each of which comprises a basic list of components: a source to provide the particles; a ring of magnets to keep the particles on a circular path; a number of accelerating stations located around the ring; and some means of injecting the particles into the ring and extracting them once they have reached the desired energy (Figure 7). Timing and synchronization are paramount in the operation of such a

Figure 7   *The basic components of a synchrotron. Particles from a source receive some preliminary acceleration in the injector before entering the main ring, where they circulate many millions of times, picking up small amounts of energy each time they pass through radiofrequency (RF) cavities. Steering magnets keep the particles on their path through the beam pipe which is maintained at high vacuum. In addition, focusing magnets (not shown) prevent the beam from spreading. Once the particles have reached the desired energy, they are ejected from the ring by an extractor magnet and directed towards experiments*

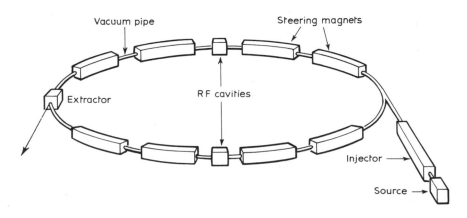

machine. At the energies necessary to unlock the atomic nucleus, the particles are travelling only fractions of a per cent less than the speed of light, or nigh on 300 million metres per second.

The importance of timing in a synchrotron is manifest in two ways. First, when the particles reach the accelerating stations, the electric field must be in the right direction – of the right polarity – to give the particles a push rather than slow them down. Because the particles are moving extremely fast, this means that the electric field must oscillate very rapidly so as to be of the right polarity each time some particles pass through. In modern accelerators frequencies up to the region of 200 million hertz (cycles per second) are typical, where the time between successive peaks of the electromagnetic wave – the time to change polarity and back again – is 5 nanoseconds (ns) or 5 thousand-millionths of a second, that is, the time for a particle moving at the speed of light to travel 1.5 metres. Moreover, the frequency must change during the course of one acceleration cycle, increasing to keep pace with the particles as their energy increases.

It is not only the electric field that must be synchronized to the increasing energy of the particles if the accelerator is to function properly. As the particles are accelerated they gain momentum and become more resistant to the influence of a magnetic field. In other words, in a constant magnetic field, such as a permanent magnet provides, a particle's path becomes less and less curved as the particle gains energy. The way to keep the accelerating particles on the same path is to increase the magnetic field accordingly, and this can be achieved using electromagnets in which the field is controlled by altering the current.

The largest modern synchrotrons are proton accelerators. The Super Proton Synchrotron at CERN, the European centre for particle physics near Geneva, has a magnet ring 2.2 km in diameter, with a total of over 1200 magnets, not only to guide the protons but also to focus the beam. The maximum accelerating push of this machine is equivalent to passing the protons through an electric field of 500 thousand million volts, to velocities 0.999998 the speed of light. The huge size of the ring is a consequence of the high energies the particles ultimately reach. There is a practical limit to the highest fields attainable in the synchrotron's electromagnets, and this means that the only way to higher energies is to opt for larger rings with less curvature, so that the magnets can keep control of particles at higher energies.

Picking up the pieces

Accelerators provide the high-energy projectiles to probe the atomic nucleus, but how do we observe the submicroscopic debris that results and how do we interpret what we find? The high-energy protons or electrons from a synchrotron can be used either directly to bombard appropriate targets, or to generate secondary beams of other kinds of particle, which are then used to investigate the target material. The simplest target from the point of view of the fundamental physics, though not from a practical point of view, is liquid hydrogen. Hydrogen being the simplest element, with one electron orbiting one 'proton, presents a nuclear target of single protons (the electrons are essentially negligible). Targets of carbon, say, or, some metal, are easier to handle, but make the analysis of the debris more complicated. The nuclei of these heavier elements contain not only protons but also neutrons, and the two types of nucleon both move around relative to each other, although bound within the atomic nucleus. Such complex targets provide different information, complementary to that from simpler targets, and so they too have an important role in our attempts to understand the nature of the nucleus and its contents.

Detecting the debris from the collisions is an art in itself, and the techniques used have changed over the past fifty years as particle physicists have pushed available technology to its very limits. While there are many varieties of particle detector, the underlying principles are common to all. The basic aim is to be able to follow the paths of whatever objects are produced in the collisions and to measure as many parameters of the objects as possible, in order to be able to identify them.

The simplest collisions are the so-called elastic collisions where, for example, an incoming proton is deflected by a target proton, and, just as in a game of subatomic billiards, the identity of the particles remains the same throughout the scattering process. But although the protons remain protons, they may, like the billiard balls, transfer energy, momentum and even their equivalent of spin, so it is important to know in which directions the protons go after the collision, how much momentum each carries, and so on.

The outgoing particles lose energy as they travel along and leave telltale signs of their passage, tracks that can be recorded and studied to reconstruct the events of the collision. Protons, and other kinds of electrically charged particle, can ionize the atoms of any substance through which they pass. That is, they knock electrons out of the atoms, leaving 'ions' with a net positive charge. If the protons pass through a gas, for example, it is possible to use an electric field to attract the electrons to one or more positive electrodes, where they will generate a pulse of electric current – a signal that can be registered by appropriate electronic circuits. In practice, the field is made sufficiently strong to accelerate the electrons enough to induce secondary ionizations. The process liberates more electrons, and so on until an 'avalanche' of electrons forms, which can produce a detectable pulse of charge.

Alternatively the protons may lose energy by exciting the atoms they pass, boosting electrons into orbits of higher energy. In materials known as scintillators, the excited atoms lose their excess energy by radiating light. The flashes of light can be picked up by sensitive detectors that absorb the light and emit an electrical signal.

A series of detectors, whether electrodes in a gas, pieces of scintillator or whatever, will indicate points along the track of a charged particle, which can later be reconstructed in analysing the data. In addition, the size of the signals produced by the passing particle often reveals the rate at which it is losing energy, and this in turn is related to the particle's energy. Slow-moving particles produce more ionization over a given distance and therefore create larger signals.

Some types of detector use the tracking medium to provide the target, a technique that has the advantage of making the tracks visible in the immediate vicinity of the collision. Much of the early work studying the high-energy collisions of cosmic rays used 'nuclear emulsion', a special type of photographic emulsion that when developed reveals the ionized trails of charged particles. Cosmic rays can simply collide with nuclei in the emulsion, and the developed emulsion will show not only the tracks of the resulting debris but also the particle initiating the interaction.

Bubble chambers are more complex detectors, which have played an extremely important role in pinpointing particles the instant they emerge from a collision. A bubble chamber contains a liquid which, on a sudden expansion of the chamber, can become superheated – that is, at the new lower pressure it remains a liquid above its normal boiling point. This superheated liquid is sufficiently unstable for the tiny amount of energy lost by a charged particle as it ionizes atoms in the liquid to initiate boiling; in other words, bubbles begin to form. Photographs, or nowadays even holograms, record the trail of bubbles along the track of the particle. And, as the nuclei of the liquid – which can be simply liquid hydrogen – provide suitable targets, the bubble chamber can record the instant of collision as well as its aftermath.

Track detectors pinpoint the particles emerging from collisions, but do not necessarily provide enough information to identify different types of particles. On their own they reveal at most the direction and energy loss of a particle. However, the addition of a magnetic field provides a handle to both the particle's electric charge and its momentum. A magnetic field bends particles of opposite charge in opposite directions. Moreover, the amount of bending depends on the particle's momentum: the same principle that keeps charged particles on their path through the magnet ring of a synchrotron allows the products of the collisions to be separated out.

Indeed, a suitable magnetic field provides the direct analogue of a prism in an optical spectrometer, only in this case the particles are dispersed according to their momentum, those of highest momentum being bent the least.

But charge and momentum are not sufficient to identify particles uniquely; for example, protons and positrons (antielectrons) are both positive, so they have the same charge, and they can have the same momentum. However, they do have widely different masses, the proton being much heavier by some 1800 times. Momentum is the product of mass and velocity, so some means of sifting out particles of the same charge and momentum according to their velocity can often provide the final clue to their identity. The rate of ionization, or energy loss, may be sufficient; a slow proton will leave behind a heavily ionized trail, whereas a positron of the same momentum but much higher velocity will leave only a light track. Another way is to look for the phenomenon of Čerenkov radiation. This is light emitted by charged particles passing through a medium – water, for instance – at a velocity faster than the speed of light *in that medium*. The light is a kind of electromagnetic shock wave, analogous in some ways to a sonic boom, and the angle at which it is emitted depends on the particle's velocity.

Detection in particle physics does not stop at identifying particles that travel some distance from the collision. We can work back from whatever is produced in the collision, even if it is a simple elastic 'billiard ball' collision, to piece together a picture of what happened during the impact itself, so as to make sense of the processes involved. The appropriate tools are no longer the pieces of highly technical apparatus, but those of calculation and the laws of physics at velocities approaching the speed of light.

In analysing the raw data from the collisions, we can use the known consequences to calculate what may have happened. There are certain universal physical laws that must hold true even in the interactions of the smallest fragments of matter. We know that energy, momentum and electric charge must be conserved. In other words, the total energy, momentum and charge must all be the same after the collision as before. Here the word 'energy' applies in its broadest sense and includes 'mass-energy', for as Einstein showed in his work on relativity, mass is merely a form of energy. Thus the conservation of energy in particle collisions must take into account any new mass, in the form of new particles, created from the available energy liberated by the processes at work in the interaction. From the known energies, momenta, directions, and so on, of the final particles, we can look for the fingerprints of other particles that maybe lived for too short a time to be registered by tracks in detectors, just as the lines in atomic spectra are the fingerprints of atoms in short-lived excited states.

The fingerprints of extremely short-lived states can show up in several ways in careful analysis of the measurements on the detected particles. The values of momenta, angles, and so on, measured in the laboratory, can be gradually moulded into a form that echoes any intermediate states. One common way to look for such states indirectly is to study the rate at which an interaction occurs with increasing energy. The reaction can be as simple as the elastic, billiard-ball scattering of two particles, but the rate at which it occurs can vary with energy, rising to peaks, or 'resonances', which imply the production of intermediate short-lived states. Alternatively, the energies and momenta of an outgoing pair of particles, say, can be used to calculate their total mass-energy. A plot of the mass obtained from the varying values of the outgoing momenta and energies may show peaks, each corresponding to the mass of an intermediate particle (Figure 8). Whichever way the 'bumps' appear on the plots, it is possible to calculate corresponding values for spin and charge, and other properties relevant only in the subatomic realm, which will be introduced in chapter 3. Thus the bumps indicate the presence of particles almost as clearly as do tracks in an ionization detector.

Together the tools of particle physics have revealed a wealth of short-lived intermediate states in addition to others observed more directly. Their study provides a 'spectroscopy' richer still than that of the atoms, and it is this spectroscopy that leads to the fundamental constituents of matter.

Figure 8 *Resonance bumps in the effective mass of annihilating electron–positron pairs reveal the existence of three members of the 'upsilon family' – particles comprising a bottom quark and its antiquark. (From T. Bohringer et al., Physical Review Letters, vol. 44 (1980), p. 1111)*

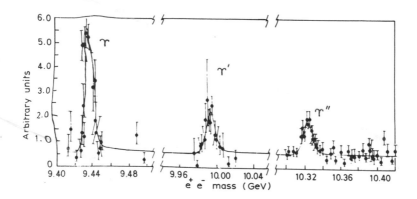

3

The Nature of Particles

Progress in our understanding of the nature of matter – as in any other scientific investigation – proceeds by two kinds of discovery: those that we expect and which appear in the right place, and those that come as a complete surprise. We can, of course, also discover that we were wrong, although such discoveries tend to become forgotten with the passing of time. But whatever the nature of the discovery, we always learn something that enhances and enlarges our view of the physical world. Experiments in atomic spectroscopy in the late nineteenth century were not done with the aim of investigating the structure of the atom; far from it, for the very concept of the atom was only just beginning to be accepted. More recently, during the past fifty years, the techniques of particle physics have exposed a new level of structure, through new kinds of spectroscopy, and through new and sometimes unexpected patterns of behaviour.

As we saw in chapter 1, the view that there are four elementary particles – the proton, neutron, electron and neutrino – with complementary antiparticles began to crumble in the very same decade that it had been established. Only five years after the discovery of the neutron in 1932, experiments tracking the interactions of cosmic rays in cloud chambers observed a particle heavier than the electron but lighter than the proton. Cloud chambers, in a sense the forerunners of bubble chambers, contain a vapour which on sudden expansion becomes supersaturated, and in which droplets of liquid form along the ionized trails of particles. The new particle of 1937 we now know as the muon, with a mass 206 times that of the electron, but with otherwise very similar properties. Its discovery marked the beginning of a new era in the study of the nature of matter.

In the 1940s and early 1950s, following the Second World War, experiments with cosmic rays in cloud chambers and in nuclear emulsions uncovered a number of new, unstable particles, all of which were visible through the tracks of the products of their decay, if not visible through their own tracks. At the same time, experiments at the new generation of particle accelerators confirmed the discoveries made with cosmic rays, and it soon became clear that there were a number of different particles, both charged and neutral. The 'new' particles decayed into other, more stable, particles, and ultimately would transmute into protons, electrons and the invisible neutrinos.

The route to understanding the new particles lay in classifying them. How were they like the known particles, how were they different? What were their masses, their electric charges, their lifetimes? One factor that soon became clear was that there often appeared to be differently charged versions of the same particle. This idea was not new. Back in 1932 Werner Heisenberg had suggested that the proton and neutron could be regarded as positively charged and neutral versions of the same particle; after all, they have very nearly the same mass, the neutron being slightly heavier by about 0.1 per cent. Moreover, experiments in nuclear physics showed that neutrons and protons respond in more or less identical fashion to the nuclear forces binding them within the nucleus; in other words, this force seems indifferent to the specific identity of the nuclear particle. So it seemed reasonable that neutron and proton were two equally likely manifestations of the same basic nucleon state. For this concept to be useful, however, it had to be compatible with the quantum theory already developed to describe subatomic behaviour. And to this end an analogy proved useful – an analogy between the nucleon states and the 'spin states' of a particle.

Spin is an intrinsic angular momentum associated with a particle; different particles have different amounts of spin, and indeed some have no spin at all. Crudely speaking, we can regard a particle with spin as a spinning top, although, as is often the case when dealing with quantum phenomena, we should not take this analogy too far. The basic idea of spin was conceived in 1925, by two young physicists at the University of Leiden, George Uhlenbeck and Samuel Goudsmit. They were seeking an explanation for the fine structure in the lines of spectra from hydrogen and alkalis such as sodium: features that at first appear as single lines in the spectra are actually seen to be composed of two lines on closer inspection. The overall spacing of the major lines in a spectrum reveals the electronic structure of the atom, so what is the significance of this finer division?

Niels Bohr had begun to explain the main lines in atomic spectra in 1913, when he proposed that the electrons in an atom could take up only certain specific orbits, characterized by particular values of energy and momentum. He was attempting to solve an important difficulty with Ernest Rutherford's simple picture of an electron orbiting a central nucleus. Electrons in circular orbits are in effect under constant acceleration, due to the influence of the central force. They should therefore radiate energy according to the dictates of electromagnetic theory. (Radiowaves, for example, are produced by electrons made to oscillate and thus accelerate in a specific way.) Why then do the electrons in an atom not radiate continuously, and spiral in to the nucleus as they lose energy?

The answer, as Bohr postulated, lies in the concept of quantization, put forward in 1901 by Max Planck. Planck had suggested that electromagnetic energy is radiated in 'lumps' or 'packets' of a certain size. The size of the packet, or 'quantum', varies according to the oscillation frequency of the electromagnetic radiation, with radiation at higher frequencies having larger quanta. Thus light, with its higher frequencies, has 'larger', more energetic, quanta, than do radiowaves with their lower frequencies.

Bohr's bold step, for which there was no real precedent, was to go a stage farther than Planck had done, and postulate that the orbits of the electrons in an atom had also to be quantized. In other words, only certain orbits where the electrons have particular values of energy and momentum are allowed. The electrons cannot have energies between the fixed values of these allowed orbits, and so they cannot radiate their energy away continuously, as electromagnetic theory alone suggests they should. The only way an electron can move from one orbit to another is by emitting or absorbing a quantum of energy. The size of the quantum is simply the difference in energy of the two orbits; the larger the difference, the higher the frequency (or equivalently the shorter the wavelength) of the emitted radiation (Figure 4).

Bohr's theory, therefore, not only overcomes the possibility that atoms radiate away to nothing, it also provides a quantitative relationship between the electron orbits and the emitted radiation, which can be checked against the wavelengths measured in atomic spectra. The original theory is somewhat oversimplified and applies only to the hydrogen atom, with one electron. With more electrons, complications set in. However, in the 1920s, quantum mechanics, developed by Erwin Schrödinger and Werner Heisenberg, provided a more accurate, mathematically sound theory of the atom, which could deal adequately with atoms with more than one electron.

So much for the main lines in atomic spectra, but what of the fine splitting that puzzled Goudsmit and Uhlenbeck? The implication is that each atomic energy level is split into two closely spaced levels; in other words, there are two closely spaced but different energies associated with each allowed orbit. What could produce such a small difference? The answer proposed by Uhlenbeck and Goudsmit was to endow the electron with an intrinsic angular momentum, or spin, in addition to its momentum from circling the

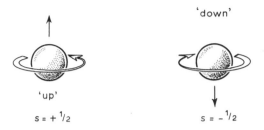

Figure 9 *We can imagine an electron as a spinning top. The 'spin arrow' points in the direction followed by a right-handed screw thread turning the same way as the particle. If the electron is spinning anticlockwise when viewed from above, we say the spin is 'up'; if spinning clockwise, the spin is 'down'. The spin quantum number, s, takes the value $+1/2$ or $-1/2$ for spins 'up' and 'down' respectively*

nucleus. Like the overall orbital angular momentum, this spin angular momentum had to be quantized, but with only two possibilities allowed, thus giving rise to the two slightly different energy levels.

To imagine these two possible states, we can think of the electron as a spinning top, able to spin either clockwise or anticlockwise about a vertical axis (Figure 9). With the electron spinning one way, the total energy of the system is slightly more than when it spins in the other direction. Thus two slightly different values of the *total* energy are associated with the main orbit, and we observe two closely spaced lines in the spectrum, rather than a single line.

This diversion into atomic physics seems to be straying a long way from the discussion of the similarity between proton and neutron, so why are the topics related? Consider first how we can represent the spin of an electron on a diagram. We can draw an arrow (or more technically, a vector) such that its shaft is aligned with the direction of the spin axis of the electron. If we imagine a right-hand corkscrew turning in the same sense as the particle is spinning, then we can define a direction for the arrow's head. The convention physicists use is for an upward arrow to represent an anticlockwise spin, and a downward arrow to show a clockwise spin. Thus we can loosely talk about spin 'up' and spin 'down'.

Now we can return to the proton and neutron at last. Just as we represent the electron's spin by two arrows in real space, pointing up and down, so we can represent the proton and neutron by two arrows in an imaginary space. The proton's arrow can point up, and the neutron's arrow down. Thus the two particles appear as different manifestations of the same basic

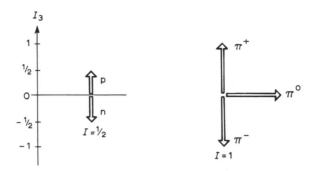

Figure 10   *The proton and neutron can be imagined as two differently charged states of a more basic nucleon, or, by analogy with spin, as two states in which an internal isospin arrow points in opposite directions. If we assign the nucleon isospin I = $^1/_2$, then the proton and neutron can correspond to isospin 'up' and 'down', or I_3 = $+^1/_2$ or $-^1/_2$, where I_3 is the value of I projected onto a vertical axis. The pion comes in three charge states, π^+, π^0, π^-. It has isospin I = 1, with three projections, I_3 = +1, 0 and −1*

object, depending only on the orientation of an intrinsic 'arrow', which by analogy with spin is known as *isospin* (Figure 10).

So, in the 1930s there emerged the idea of a new property, isospin, with which to describe subatomic particles (and, in fact, nuclei composed of those particles). Like spin, but unlike electric charge, this new property has no direct manifestation in the macroscopic world; it belongs only to the quantum world within the atom. However, it provides another useful label with which to classify particles.

Strange particles: classification by quantum number

Labelling particles according to charge, spin, isospin, and so on, is in more technical terms the process of ascribing quantum numbers to them. The label is the quantum number, a concept again derived from atomic physics. In Bohr's simple picture of an atom, each orbit is defined by a whole number, denoted n. Bohr postulated that the angular momentum (as well as the energy) of each orbit had to be quantized, that is, allowed only certain values. These allowed values are given by the expression $nh/2\pi$, where n is an integer and h is a constant,* which determines the basic size of a quantum. The orbit of lowest angular momentum and lowest energy –

---

* Planck's constant.

the orbit nearest the nucleus – corresponds to a value of n equal to 1; the next orbit has n equal to 2, and so on. In this way the value of n describes one unique state out of many possible electron orbits. The quantum number n is a label which defines a particular atomic state.

In a similar way, we can use a quantum number, this time denoted by s, to define the spin of a particle and so give it a label. Goudsmit and Uhlenbeck found that the electron must have spin $\frac{1}{2}h/2\pi$, so we can label the electron with the quantum number $s = \frac{1}{2}$. Now we can make the electron's spin arrow half a unit long, so that the length of the arrow indicates the size of spin. And we can go farther, and associate a sign with s depending on the direction of the spin arrow: for spin up, we say $s = +\frac{1}{2}$, for spin down $s = -\frac{1}{2}$. Similarly, we can define a quantum number, I, to label a particle with a tag describing its isospin. By analogy with the two spin states of the electron, the nucleon with two isospin states (the proton and neutron) is given the label $I = \frac{1}{2}$. The so-called 'third component of isospin', I_3, labels the individual states. For the proton, with the isospin 'arrow' up, $I_3 = +\frac{1}{2}$; for the neutron, $I_3 = -\frac{1}{2}$ (Figure 10). A useful shorthand is to refer to the electron as having spin $\frac{1}{2}$, and the nucleon as having isospin $\frac{1}{2}$.

Thus, as new particles began to reveal themselves in experiments in the late 1940s, there were available several labels to help in classifying the new discoveries. Along with mass and lifetime, which appear to be unquantized and can take almost any value, there were charge, spin and isospin, all represented by quantum numbers. (Note that the size of the electron's charge of 1.6×10^{-19} coulombs, often denoted by e, seems to be a fundamental unit of charge, and that all observed particles have charge $\pm Qe$, where Q is a quantum number equal to 0, 1, 2 . . . and so on. So charge, like spin, is apparently quantized.)

Among the first particles to be discovered in the cosmic-ray experiments of the late 1940s, were the pi-mesons, or pions. The existence of such particles had in fact been predicted by the Japanese theorist Hideki Yukawa in 1935, in his theory of the strong nuclear force which binds protons with neutrons within the nucleus. Yukawa envisaged the pions as carrying the force between the nucleons in the nucleus, as chapter 4 explains in more detail. He predicted that there should be a particle with a mass about a tenth that of the proton, in other words, between that of the electron and the proton – hence the name 'meson' from the Greek for 'middle'. The pion was found at last in 1947, when a team from the University of Bristol studying nuclear emulsions exposed at the French observatory on the Pic du Midi in the Pyrenees discovered short tracks made by pions before they decayed into the particles known as muons.

The pion turns out to exist in three charge states: positive, neutral and negative. So how does it fit into the isospin scheme that works for the nucleons? The answer is that in terms of isospin the pion is analogous to a particle with spin 1 (or spin $h/2\pi$, to be more precise). In general, the number of spin states for a particle with spin s is $2s + 1$. So for the electron with spin $1/2$, there are two states, 'up' and 'down'. A particle of spin 1 can have three quantized states, which we can represent on a diagram as 'up', 'down' and 'sideways'. The states are labelled by the quantum numbers $s = +1$, -1 and 0 respectively. As for the pion, its three charge states correspond to three states in isospin space. Thus we say the pion has $I = 1$, with $I_3 = +1$, 0 and -1 corresponding to the three states observed (Figure 10).

The concept of being able to label particles not only according to their mass but also in terms of quantum numbers such as s (for spin) and I (for isospin) has a profound importance. Just how important did not become clear until many more particles had been discovered and classified, and new kinds of label had been found necessary.

In the years following the first observations of the pion, experiments revealed more new particles, some heavier than the nucleons, others with a mass between that of the pion and the proton. These new particles behave in ways that were at first difficult to understand. In particular, while they are produced relatively copiously, being about a tenth as common as pions, they decay relatively slowly, into pions and nucleons, within a timescale of 10^{-8} to 10^{-10} seconds, or a few thousand-millionths of a second. Although this does not sound a long time, it is very long when compared with the time for strong nuclear interactions. These take place in the time it takes light to cross a nucleus, roughly 10^{-23} seconds. Pions and the other new particles should be created in strong nuclear interactions in cosmic-ray collisions on this sort of timescale; why then should they not decay at the same rate? The pion decays into a particle that does not take part in strong nuclear interactions – the muon – and the slow decay of the pion is thought to be a weak interaction like the decay of the neutron. But other new particles decay into pions and protons, which can take part in strong interactions, so why should their decays be so slow?

The answer to the paradox lies in yet another label – another quantum number – this time rather mischievously called strangeness, in accordance with the apparently strange behaviour of the new particles. In 1952 Abraham Pais at the Institute for Advanced Study in Princeton introduced the idea of 'associated production'. The new particles were observed to come in two classes: those that are lighter than the proton, the so-called K mesons, or kaons; and those heavier than the proton, which are known collectively as hyperons, from the Greek for 'over'. Pais hypothesized that kaons and hyperons are always produced in pairs: if a hyperon is produced, then so must be a kaon. But in the decay of a kaon or hyperon, the particle is 'on its own', as it were. Pais's conjecture was that the associated production of kaons and hyperons was the manifestation of a rule that had to be obeyed in the *strong* interactions that formed the particles. The decays observed did *not* obey the rule, and therefore they must be due to *weak* interactions, as the timescale of their occurrence indeed suggested.

Experiments confirmed that Pais was right, at least as far as they could, in that no one found an example where associated production did not hold. But it was theorists Murray Gell-Mann from the California Institute of Technology (Caltech) and Kazuhito Nishijima from Osaka City University who independently put the observations on a more formal footing. They suggested that the phenomenon of associated production reflected the conservation of some property, with its own quantum number, which they called 'strangeness'. Just as electric charge must be conserved, as when a neutral neutron decays into a positive proton and a negative electron, then so must strangeness be conserved, but only in strong interactions. If a positive particle is created in a nuclear interaction, then so must be a negative particle; likewise 'strange' particles can be produced only in pairs of overall zero strangeness.

How does this work in practice? One common example of associated production occurs when a negative pion (π^-) interacts with a proton (p) to produce a neutral kaon (K^0) and the lightest hyperon, which is known as the lambda (Λ^0) and which is also neutral. The reaction can be written as follows:

$$\pi^- + p \rightarrow K^0 + \Lambda^0.$$

Electric charge is conserved, being a total of zero on either side of the equation. Strangeness is conserved too if we give the Λ^0 a strangeness of -1 and the K^0 a strangeness of $+1$. (The pion and proton each have zero strangeness.) In their decays by the weak interaction, the Λ^0 transmutes into a proton and a π^-, the K^0 into a $\pi^-\pi^+$ pair, these processes involving a change of strangeness of $+1$ and -1 respectively.

The Gell-Mann–Nishijima scheme was by no means as arbitrary as this might suggest. They in fact developed a formula that interrelated quantum numbers known to describe a particle: electric charge (Q), isospin (I), strangeness (S) and baryon number (B). The baryon number expresses to a certain extent the stability of matter in the universe. The early research on hyperons showed that they always decayed finally to a proton, as does the neutron. And the proton seemed to be stable. Thus arose the idea that the heavy particles must carry a property that is conserved, again just as electric charge is.

The heavy particles – the hyperons and the two nucleons – became known as baryons, from the Greek for 'heavy', and they were assumed to carry a quantum number called the baryon number, B. In any interaction the total baryon number must remain constant, and as the proton is the lightest baryon, all others must decay ultimately to a proton, which can decay no farther without changing baryon number. The hyperons and nucleons each have a baryon number $B = 1$, while for the antibaryons $B = -1$. On the other hand, the mesons – the pion and kaon – are not baryons and they are labelled with $B = 0$.

Gell-Mann and Nishijima related S and B with the other quantum numbers Q and I in a simple formula, the beauty of which is that it ties conservation of strangeness in with conservation of baryon number, isospin and electric charge. The rule that strangeness cannot change in strong interactions no longer seems so arbitrary. In this way the Gell-Mann–Nishijima formula brought order to the relationship between the new strange particles, and the other, more familiar, particles. It even predicted the existence of some particles before they were discovered, but that is another story.

Quarks: classification by symmetry

Strangeness gives us a new label which helps us to classify particles neatly and provides us with rules for what may or may not take place in particle interactions. But it barely helps in developing a theory of particle interactions, which we need if we are to understand what takes place, and to predict the rate at which reactions occur and the time it takes particles to decay. Perhaps most important of all, it leaves open the question of why some particles are endowed with strangeness; and indeed why there are so many kinds of particle. The process of classification is vital to further understanding, however, and the particle physicists of the 1950s and early 1960s were in many ways like the scientists of the latter part of the nineteenth century who studied the detailed structure of atomic spectra, little realizing that they were providing the observational groundwork for the quantum theory of the atom that was to develop from Bohr's ideas in the 1920s.

The pattern that began to emerge in the 1950s was that baryons (or mesons) of more or less the same mass could be grouped in multiplets (singlets, doublets and triplets) of isospin, each member having a different electric charge. But in the early 1960s a number of theorists, including Gell-Mann and the Israeli physicist Yuval Ne'eman, began to develop larger patterns comprising more particles. Suppose, for example, you plot a graph with strangeness as the vertical axis and isospin, specifically I_3, as the horizontal axis (Figure 11). You can assign the lightest known baryons (including the proton, p, and the neutron, n) to points on the

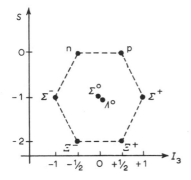

Figure 11 *If we plot strangeness, S, along the vertical axis, and the projected value of isospin, I₃, along the horizontal axis, the eight lightest baryons lie on a hexagon, with two particles at the centre. Note that these baryons all have a spin of* $1/2$

graph with the corresponding values of S and I_3. The eight particles, denoted by the symbols p, n, Λ^0, Σ^+, Σ^0, Σ^-, Ξ^-, and Ξ^0, form a symmetric hexagon shape with two particles, the Λ^0 and Σ^0, at the centre. (The Σ and Ξ particles are hyperons, heavier still than the Λ.) It is worth noting here that these baryons all have spin $1/2$, so in a sense they appear as different manifestations of the same basic baryon with spin $1/2$. What interested Gell-Mann, Ne'eman and the others was that the symmetry of the baryon 'octet' opened the way to using the mathematics of symmetry groups to work out transformations between particles, such as decays from one sort to another.

A symmetry group contains all the changes you can make to an object while retaining its symmetry. A snowflake, like the baryon octet, has hexagonal symmetry: it can be rotated as many times as you like through an angle of 60° and will always look the same. Symmetry is important in physics because it is deeply connected with conservation laws such as those for energy and electric charge, and isospin and strangeness. If a system is symmetric, like the snowflake, then when changed in a certain way it will appear the same, as does the snowflake when rotated through 60°. Such invariance implies conservation: everything stays the same. If a physical law is invariant under a symmetry operation like rotation, then some quantity is conserved. Indeed, the fact that the results of an experiment are the same if it is turned through 60° reflects the conservation of angular momentum. And the discovery that neutrons and protons respond in the same way to the nuclear force binding them in the nucleus reflects the conservation of isospin in these strong nuclear reactions.

The symmetry that Gell-Mann and Ne'eman independently investigated for describing the elementary particles is called SU(3) and it provides rules that connect the particles within multiplets such as the baryon octet. SU(3) stands for 'special unitary group in three dimensions'; its basic 'irreducible representations', analogous to the six possible positions of the snowflake under rotational symmetry fall into groups of one, eight or ten. Suppose we identify the octet of baryons with one of the groups of eight representations of SU(3). Is there another group of particles we can identify as a 'ten', or decuplet?

The answer is yes, provided we turn to the so-called resonances. The baryons and mesons discovered in the emulsions, cloud chambers and bubble chambers of the 1950s distinguished themselves by their relatively long lifetimes, which in turn showed that these particles decayed not by the strong interactions that produced them but by the weak interaction responsible for neutron decay. However, a different class of experiments at accelerators soon revealed particles that decay by strong interactions, in other words, on timescales of roughly 10^{-23} seconds. This is far too short a time for a particle to leave a discernible track in a detector, but the very fact that a new particle has been formed before decaying does influence the outgoing particles. Study of the decay products in the appropriate way reveals peaks, or maxima, in the numbers of particles produced at certain energies – resonances. And by careful labelling of the ingoing and outgoing particles, it is possible to assign definite properties, such as spin, isospin and strangeness, to the intermediate resonant state.

Some of the first resonances discovered appeared in data for the scattering of energetic pions off nuclei, in other words protons and neutrons. As the pion has isospin 1, and the nucleon isospin $1/2$, they can, according to the rules of spin combination, form particles of isospin $1/2$ or $3/2$. Those resonances with $I = 1/2$ became known as N resonances, those with $I = 3/2$ as 'deltas', or Δ. Consider the delta for a moment: with $I = 3/2$, it has $2I + 1 = 4$ possible states – it is a quadruplet in isospin space; and it has spin $s = 3/2$. Other resonances with spin $3/2$ are the triplet known as the Σ^* resonances (observed in kaon–nucleon scattering) and the doublet known as the Ξ^* resonances. These resonance particles can be plotted on a grid of strangeness versus isospin, just as the spin-$1/2$ baryons were, and they reveal an interesting pattern (Figure 12). There are nine states in all, and the pattern is very suggestive of a tenth state: a triply strange ($S = -3$), negatively charged isospin singlet ($I = 0$) of spin $3/2$. Such a particle was found early in 1964, and could be seen to decay through three stages, shedding one unit of strangeness at each step. The discovery of this particle, known as the Ω^-, was great vindication for the investigators of SU(3) and showed that the symmetry did indeed seem to have some important foundation in the natural world.

SU(3) alone is but a symmetry scheme; it does not provide any clues as to why particles should behave according to the rules of this particular mathematical group. But in 1964 the success of SU(3) led Gell-Mann and a fellow physicist from Caltech, George Zweig, who was working for a year at CERN, to arrive independently at a more fundamental conclu-

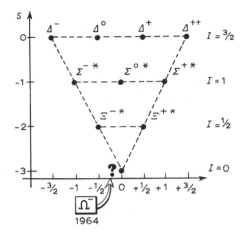

Figure 12 *On the strangeness/isospin grid, the lightest spin-3/2 resonances known in 1961 formed a pattern like an inverted triangle, although a particle to occupy the position at the vertex was missing. The Ω⁻ particle, triply strange with S = −3, was discovered in 1964, and neatly fills the space*

sion. The representations of SU(3) – the octets and decuplets – can all be generated from a basic triplet, which at first seems to have no direct analogue in nature. The innovative suggestion of Gell-Mann and Zweig was that the triplet corresponds to three 'fundamental' particles, from which all the others could be built according to the rules of SU(3). Gell-Mann called the hypothetical particles 'quarks', and the name has stuck. They have been given the individual names 'up', 'down', and 'strange', or u, d, s for short, and, once assigned appropriate quantum numbers, they can be used to construct most of the particles known in 1964.

The quantum numbers of the quarks turn out to be rather unusual (Table 1). The baryon multiplets can be generated if the members of the basic triplet – the quarks – are brought together in threes. With three quarks to a baryon, the baryon number of a single quark has to be 1/3. Even more surprising, the charges of the quarks must be fractions of the basic unit of charge, *e*, the size of the electron's charge. And to account

Table 1 *Quantum numbers of the three quarks proposed by Gell-Mann and Zweig in 1964. These three quarks were sufficient to form all the strongly interacting particles (hadrons) known at the time*

Quark type	Baryon number B	Charge Q	Isospin I	I_3	Strangeness S	Spin s
Up (u)	$1/3$	$+2/3$	$1/2$	$+1/2$	0	$1/2$
Down (d)	$1/3$	$-1/3$	$1/2$	$-1/2$	0	$1/2$
Strange (s)	$1/3$	$-1/3$	0	0	-1	$1/2$

for the strange particles, the strange quark has $S = -1$, while this quantum number is 0 for the other two quarks. Lastly, the quarks must have spin $1/2$, so that the three quarks in a baryon can align their spin arrows with two parallel and one antiparallel to give a total spin $1/2 + 1/2 - 1/2 = 1/2$; or they can all be parallel to give $1/2 + 1/2 + 1/2 = 3/2$. Using the rules of SU(3), we can then build up the baryon octet and decuplet from the basic triplet and so discover the quark construction of the various particles. The proton and neutron contain 'uud' and 'udd' respectively; the Ω^- is 'sss' – it is triply strange because it contains three strange quarks (Figure 13a).

How do the mesons fit into this picture? Just as the lightest baryons with spin $1/2$ correspond to an octet of SU(3), so do the lightest mesons with spin 0, such as the π and K mesons. They too can be plotted on the grid of strangeness versus isospin, where the particle that shares the centre spot with the π^0 is the eta meson, η^0, a resonance that decays into two pions (Figure 13b). The mesons thus fit neatly into the SU(3) symmetry, so they too must be constructed from the basic triplet of quarks. In this case, a meson octet can be generated if we consider particles built from a quark together with an antiquark – for the quarks like all other particles are presumed to have corresponding antiparticles. With spins antiparallel, the quark–antiquark pairs form the spin-0 mesons; with parallel spins, the pairs form other mesons, with spin 1. Thus the π^+ contains u$\bar{\text{d}}$, where the bar indicates an antiquark; and the K$^+$ is u$\bar{\text{s}}$, with strangeness $S = 1$, consistent with the prescription of Gell-Mann and Nishijima (Figure 13b).

Figure 13   *From a basic triplet of quarks, u, d, s, with the quantum numbers assigned in Table 1, we can construct the observed particles. Groups of three quarks can be identified with baryons, as in the decuplet for spin $3/2$ (a). Combinations of quarks and antiquarks (signified by a bar over the letter) form the mesons, as in the octet for spin 0 (b). The π^0 and η^0 are rather more complex mixtures of quarks and antiquarks*

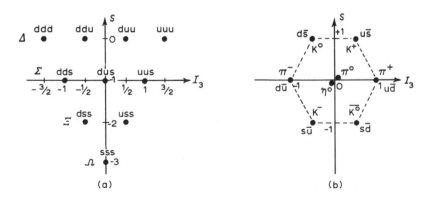

The quarks of Gell-Mann and Zweig could, of course, be mere mathematical artefacts, but evidence from many experiments since the late 1960s indicates that this is not so. In particular, experiments that collide electrons with protons or protons with protons at very high energies have indicated that the proton must have some structure; the data cannot be explained if we treat the proton as a simple point, as we can do for the electron. The proton instead seems grainy, and some parts of it seem harder than others. In other words, the proton – and the neutron – appears to be composed of smaller objects, and *all the indications are* that these objects are indeed quarks, with the quantum numbers Gell-Mann and Zweig assigned them in the SU(3) model. The italics express an important point: no experiment has yet revealed a quark directly, by its track in a bubble chamber, say. But the theory of the strong nuclear force which keeps quarks bound within protons, and protons within nuclei, suggests that it may be impossible for a single quark to have its own separate existence. The strong force apparently prevents us from seeing directly what may well be the ultimate constituents of matter.

New quarks, new symmetries

The quark model of the 1980s is quite different in some respects from that put forward by Gell-Mann and Zweig, although the basic elements remain the same. We now know that there are probably six types of quark, rather than three. In the mid-1970s a number of experiments revealed some new particles which behave like mesons but which are several times heavier than the proton. The first to be discovered has become known as the J/Ψ (pronounced 'jay-sigh') – a rather unsatisfactory compromise between the names adopted by the two groups of researchers who independently discovered the particle as a resonance in the energy of electron–positron pairs. An unusual feature of the J/Ψ is that it lives for a relatively long time, despite its being heavy enough to decay in many ways. The particle's apparent reluctance to decay is reminiscent of the slow decays of the hyperons, which gave them the name of 'strange particles' for perpetuity. However, it is the neutral meson known as the φ that provides the clue to the real nature of the J/Ψ.

The φ fits into the quark model as one of the two mesons at the centre spot of the hexagon for the spin-1 mesons, and consists of a strange quark with its antiquark (s$\bar{\text{s}}$). Thus although the φ contains a strange quark ($S = -1$), its strangeness is neutralized by the presence of the strange antiquark ($S = 1$), and the φ appears as a non-strange particle. But the φ behaves oddly in that although it should be able to decay to three pions, it prefers to decay to two kaons, a process that is energetically less

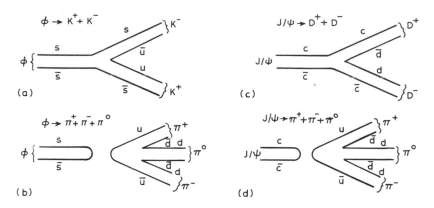

Figure 14 *These diagrams represent quarks and antiquarks by lines. Note that if a quark line reverses direction it becomes an antiquark (and vice versa); thus a loop represents a quark and antiquark of the same type – u, d, s, or c. Although it should be easier for a φ meson to decay to three pions than to two kaons, the opposite is true. The Zweig rule gives this observation the status of a law, and says that only decays where the lines connect across the diagrams are allowed. Thus the diagram (a) is favoured over (b). In the case of the J/ψ, conservation of energy prevents the process with the connected diagram (c); only the 'forbidden' process (d) can occur, so that the J/ψ lives for a relatively long time*

favourable. Zweig formulated an explanation for this behaviour in terms of a rule which has to do with the quarks the particles contain. Zweig's rule suppresses decays into particles containing entirely new types of quarks. On a diagram where lines represent quarks (and antiquarks), a decay allowed by the Zweig rule contains lines that flow from one side of the diagram to the other (Figure 14a); in a 'Zweig-forbidden' decay the lines on either side of the diagram are unconnected (Figure 14b). In the case of the φ, the s and s̄ appear in the K⁻(sū) and K⁺(s̄u) decay products, but not in the decay to π⁺π⁰π⁻.

The explanation of the J/Ψ is that it resembles the φ but contains a new type of quark and antiquark. It turns out that conservation of energy prevents the J/Ψ from decaying into the analogue of K mesons, with a diagram of connected quark lines (Figure 14c). The Zweig rule on the other hand suppresses its decay to pions (Figure 14d), so the J/Ψ ends up with a remarkably long life – and that long life is the hallmark of a new type of quark, c, carrying a new quantum number, called charm.

The existence of the 'charmed' quark was not entirely unexpected, as we shall see in chapter 6, but its discovery still came as a surprise to experimentalists. They were perhaps better prepared a few years later when heavier mesons, again unusually long-lived, turned up in experiments observing muon pairs (μ⁺μ⁻). These 'Y particles' were soon interpreted as containing quark–antiquark pairs of yet another type, b, with yet another quantum number, dubbed 'bottom'.

Since the discoveries of the J/Ψ (c̄c) and Υ (bb̄) particles, experiments have also revealed a number of particles – baryons and mesons – containing only one charmed or bottom quark in combination with the more familiar varieties. These particles are all relatively heavy, because the c and b quarks are much heavier than the u, d and s quarks, and so are difficult to form and observe in experiments. However, it is clear that there is a host of heavy, short-lived particles that shed their charm or bottom quantum numbers through decay chains just as the strange particles rid themselves of strangeness. These heavy particles fit into symmetry multiplets, analogous to those of the original SU(3) scheme, but because there are more basic objects (more quarks) to choose from, there are many more possibilities. However, a more significant symmetry group, as chapter 6 explains, turns out to be SU(2), which is based on doublets, rather than triplets, of quarks. This scheme pairs u with d, and c with s, and b with a sixth type of quark, t, for top.*

Thus in the mid-1980s we have come to the position where we believe the basic constituents of matter to be the quarks, which come in six varieties, or 'flavours', to use the 'technical' term (Table 2). Baryons are particles comprised of three quarks (qqq), while mesons contain a quark–antiquark pair (qq̄). The distinction in masses from which the names originated has become completely blurred, as we now know of

Table 2 *Quantum numbers of the quarks believed to exist in 1983*

Quark type	Baryon number	Charge	Strangeness	Charm	Bottom	Top
Up (u)	$1/3$	$+2/3$	0	0	0	0
Down (d)	$1/3$	$-1/3$	0	0	0	0
Strange (s)	$1/3$	$-1/3$	-1	0	0	0
Charm (c)	$1/3$	$+2/3$	0	1	0	0
Bottom (b)	$1/3$	$-1/3$	0	0	-1	0
Top (t)	$1/3$	$+2/3$	0	0	0	1

* At the time of writing (July 1984), the first evidence for a sixth quark, 'top', has recently emerged. The top quark must be much heavier than the bottom quark, otherwise it would have formed particles observable in many recent experiments. The proton–antiproton collider at CERN, described in chapter 8, should have sufficient energy to create particles containing top quarks; a result one way or another awaits a full analysis of the data collected at the machine has yielded the first hints of the top quark.

many mesons that are heavier than the proton. Together, the baryons and mesons form a class of particles known as hadrons, from the Greek for 'strong', for these are the particles that feel the strong nuclear force – the force that acts between quarks and quarks alone.

The lepton symmetry

There remains one other group of elementary particles, of which the electron is the most familiar member. The electron is not built from quarks. Indeed, unlike the proton and neutron, it behaves as a point as far as we can tell – that is, it is less that 10^{-17} cm in size, according to our most precise experiments. This is one thousand-millionth as small as an atom, and one ten-thousandth the size of a proton. The other members of the electron's family of particles are likewise not built from quarks, and therefore, just like the electron, are not involved in strong nuclear interactions. However, these particles do take part in weak nuclear inter-actions, as typified by the decay of the neutron.

The neutron decays into a proton, electron and a neutral particle, the neutrino (v). Although it was not discovered until 1956, the neutrino had been 'invented' in 1930 by Wolfgang Pauli, to explain apparently missing energy in observations of neutron decay. Another weak decay process is that which ends the life of a pion. Now, although a meson is built from quarks, it has a baryon number of 0, so it can decay into particles not built from quarks. What happens, in a crude sense, is that the quark and antiquark of the meson can 'cancel' each other out, so the 'quarkiness' of the meson does not have to be carried through into its decay products.

The charged pions do not decay directly into electrons however; they instead prefer to decay into muons (μ). These are particles that were first observed in cloud chambers in the late 1930s, and so were the first of the 'new' particles to be discovered. However, the muon has been difficult to understand. Although it is unlike the electron in that it is unstable and decays into an electron (or positron, depending on its charge), the muon behaves like nothing so much as a heavy electron, with a mass 206 times greater. And like the electron, the muon has its own variety of neutrino, which accompanies its interactions. Indeed, it appears that these particles all bear some special quantum numbers, rather as the different quarks have their 'flavour' quantum numbers – u, d, s, c, b, t.

Together, the muon, electron and the two types of neutrino – and their respective antiparticles – are known as leptons, originally from the Greek for 'slender' as they seemed to be the lightest of particles. Now, however, the term is taken to refer to particles not built from quarks, in other words, those that do not feel the strong nuclear force. All interactions between particles appear to have a fundamental respect for leptons, as they do for baryons, so that the total number of leptons (minus antileptons) always remains the same. Again, this conservation law is reflected in the stability of matter, and the existence of atoms.

So, we can allocate particles a lepton number, L, which is +1 for leptons, −1 for antileptons and 0 for baryons and mesons. Unfortunately, nature is rather more complicated, as it turns out that a pion cannot decay into a muon and an 'electron–neutrino' of the sort produced when a neutron decays. We have to assign quantum numbers L_e and L_μ to specify precisely what type of lepton we are dealing with. Thus the electron has $L_e = 1$, $L_\mu = 0$, the muon $L_\mu = 1$, $L_e = 0$, and so on. So when a μ^- decays to an electron, two more particles are produced:

$$\mu^- \rightarrow e^- + \bar{\nu}_e + \nu_\mu.$$

The electron–antineutrino ($\bar{\nu}_e$ with $L_e = -1$) and the muon–neutrino (ν_μ with $L_\mu = 1$) together ensure that the lepton quantum numbers are conserved.

Different types of lepton? Does this begin to sound rather familiar? Indeed, in line with the quark model, the different varieties of lepton are now referred to as different 'flavours', and as with the quarks the leptons fit into doublets of an SU(2) symmetry scheme – e with ν_e and μ with ν_μ. But this is not quite all; experiments in the late 1970s discovered a third, still heavier, charged lepton, called the tau (τ), which is nearly twice as heavy as a proton, thus showing the name lepton to be a definite misnomer. The tau has its associated quantum number, L_τ, and its associated neutrino, ν_τ, although this has yet to be observed in experiments. Together they complete a picture of six leptons to complement the six quarks (Table 3). From four fundamental particles we have progressed

Table 3 *Quantum numbers of the six leptons. Although the neutrino (ν_τ) associated with the tau (τ) has not yet been observed through its reactions, it seems reasonable to assume that it must exist in analogy with the other lepton pairs*

Lepton type	Charge	L_e	L_μ	L_τ
Electron (e)	−1	1	0	0
Neutrino (ν_e)	0	1	0	0
Muon (μ)	−1	0	1	0
Neutrino (ν_μ)	0	0	1	0
Tau (τ)	−1	0	0	1
Neutrino (ν_τ)	0	0	0	1

through myriads to arrive back at two sets of six (Figure 15). Perhaps nature is not so complicated after all.

Charge

$$\text{Quarks}: \quad \begin{pmatrix} u \\ d \end{pmatrix} \quad \begin{pmatrix} c \\ s \end{pmatrix} \quad \begin{pmatrix} t \\ b \end{pmatrix} \quad \cdot \quad \cdot \quad \cdot ? \quad \begin{pmatrix} +\,^2/_3 \\ -\,^1/_3 \end{pmatrix}$$

$$\text{Leptons}: \quad \begin{pmatrix} \nu_e \\ e \end{pmatrix} \quad \begin{pmatrix} \nu_\mu \\ \mu \end{pmatrix} \quad \begin{pmatrix} \nu_\tau \\ \tau \end{pmatrix} \quad \cdot \quad \cdot \quad \cdot ? \quad \begin{pmatrix} 0 \\ -1 \end{pmatrix}$$

Figure 15 *Six quarks, six leptons: are these all the elementary building blocks necessary to make the universe the way it is? By pairing the quarks and leptons as shown, we can perceive a symmetry that will perhaps lead us to the answer to this and other questions about the fundamental nature of matter*

4

The Nature of Forces

There is some comfort in the idea that nature is basically simple. On one level, we can appreciate simplicity because it reduces the amount we have to know and comprehend; thanks to Newton's insight, the same equation describes the motion of the planets and the path of a cricket ball. On a more fundamental level, simplicity in the physical world implies that by understanding one phenomenon we may come to comprehend a wider field, for events and properties that at first seem quite disparate may ultimately be linked by some underlying connection. In his work on electromagnetism, James Clerk Maxwell not only gained insight as to the nature of light; he also laid the foundation for our present understanding of the physical world in terms of elementary particles and the forces that hold them together. In discovering simplicity he brought us to a deeper level of comprehension.

Unfortunately, however, the world view that derives from Maxwell's work turns upside down any comfortable notions we may have of a world built with varying degrees of complexity from a basic set of building bricks. It may be difficult to imagine something as small as a quark, but the simple idea that we can proceed in stages, beginning by putting together quarks, to build something like a grain of sand or a drop of water or a protein molecule, has a satisfying feel. We can imagine the process even if we are not equipped to do the relevant calculations, and that is some progress towards comprehension. So the suggestion that the building blocks are not all that important, rather it is the nature of the forces that makes our universe the way it is, may come as rather a shock. But the reward is that, in coming to terms with the nature of the forces, we come to an understanding of the nature of particles that is more fundamental than a simple picture of building bricks.

Field or force?

What is a force? Newton sidestepped the question. He made no attempt to explain how his universal force of attraction between bodies operated, but concentrated on writing down the appropriate equations to describe the motion, which he could show to be correct by comparing theory with observation. His version of gravity became accepted as a form of 'action at a distance', in which one body is influenced instantaneously by another, without any apparent means of communication. There was no question about *how* the force of gravity makes itself felt. It was not until nearly 200 years later, in the first half of the nineteenth century, that our present ideas concerning forces began to take root, not in the realm of moving bodies but in the interwoven effects of electricity and magnetism; in the work of Maxwell, the university don, and Michael Faraday, the bookbinder's apprentice who became one of the great experimental physicists of all time.

Coulomb had shown in 1785 that the force between two electrically charged bodies depends on the product of their charges divided by the square of the distance between them. What does Coulomb's law reveal about the electrostatic force? Suppose we have an electrically charged object, which I shall call the 'source', sitting at some point in space, and that we equip ourselves with a 'test charge' with a value of one unit. At any distance from the source, the electrostatic force, which depends on the product of the charges, will be proportional to the size of the source charge, because the value of the test charge is '1'. The force will also be inversely proportional to the square of the distance between the charges. Moreover, the direction of the force will be along the straight line connecting the test charge to the source, as Coulomb's law also tells us. Thus the force on the test charge reflects directly the varying electrostatic influence of the source as we move the test charge around. We can imagine measuring the force on the test charge as we move it, plotting the values on a graph to create a map of the force's strength and direction at a variety of points in space (Figure 16).

Faraday's innovative step was to consider such a map as more than a useful computational device and to see it as representing a physical field of force emanating from the static source charge to fill the surrounding space. The field is strongest closest to the charge and decreases inversely with the square of the distance as we move away. The strength of the field at any point in space gives the strength of the electrostatic force there due to the source charge. Thus we can conceive of a second charge interacting not directly across space with the first charge but with the field arising from the first charge.

Faraday's belief in the reality of electric and magnetic fields went against

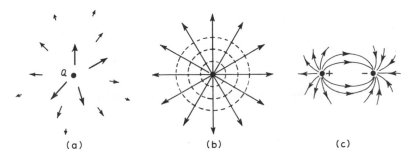

Figure 16 *We can measure the size and direction of the force on a test charge of +1 due to a charge of +Q, and represent our results by arrows at the places where we make the measurements (a). Notice how the size of the force becomes smaller farther from the charge, but always points radially outwards from the charge Q. A formal map of the electrostatic field around a point charge (b) shows lines emanating from the charge, which represent the direction of the force, and dotted concentric circles through points where the force is of equal magnitude. With opposite charges (c), electrostatic field lines flow from positive to negative charge, mapping out the force field between them*

the contemporary conventional wisdom, but fortunately Maxwell began to look more seriously at these ideas in the mid-1850s, and a decade later he had produced the electromagnetic-field theory that is the basis of much of modern physics. By writing down general equations to describe the effects of both static and varying electric and magnetic fields, Maxwell arrived at two equations describing the wave motion of electric and magnetic disturbances propagating through free space. The velocity of these waves appeared in terms of electrical and magnetic parameters that were known from measurements, and it turned out to be extremely close to the velocity measured for light. So, Maxwell concluded, light must be 'an electromagnetic disturbance in the form of waves propagated through the electromagnetic field according to electromagnetic laws.' In developing Faraday's concept of a field of force, Maxwell had shown that electric and magnetic effects do not make themselves felt instantaneously, but that they travel at a certain velocity; in free space that velocity is the velocity of light.

Light, in Maxwell's theory, is transmitted as small vibrations in an electric field, which generate a vibrating magnetic field, which in turn produces a varying electric field, and so on across space. With wavelengths between 400 and 700 nm, visible light corresponds to only a narrow part of the possible spectrum, which we can now observe over a range from less than 10^{-12} m to more than 10^6 m from gamma rays to radiowaves. The electromagnetic theory, however, left open the question of *how* the waves could travel through free space, that is, through a vacuum. Ripples on a pond, for example, are vibrations of the water, and sound waves are vibrations – small changes in pressure – in air (or any gas for that matter). But how can electromagnetic waves cross empty space, as they apparently do?

Many physicists in the late nineteenth century considered the possibility of an 'all-pervading ether', a medium through which light would propagate and which would exist even in a vacuum. The velocity of light would, they argued, be constant with respect to the ether, and not with respect to the earth, which was assumed to be moving through the ether. But in 1887 experiments by Albert Michelson and Edward Morley in the USA showed that the measured velocity of light remains the same irrespective of its direction with respect to the earth's motion through space, and therefore irrespective of the direction of flow of the hypothetical ether. Thus it seemed that empty space was truly empty, which left the propagation of electromagnetic waves through a vacuum as something of a paradox. And as so often happens in the course of physics, there were further clouds on the horizon to dull what had at first seemed to be the dawning of a bright new age where everything was clearly perceived and understood.

The quantum connection

At the turn of the century Max Planck, a professor of physics in Berlin, made a discovery that was destined to have a profound influence on the future of physical thinking. Indeed, his work marks the birth of what we now call 'modern' physics, to distinguish it from the 'classical' physics of Newton and Maxwell. Planck was concerned with solving a problem that had arisen in the study of radiation. Existing theoretical ideas could not explain the way that the intensity of radiation emitted by a hot body varies with wavelength. A glowing lump of hot coal, for example, emits over a range of wavelengths including infrared radiation, which we feel as heat. The maximum intensity is in the red part of the visible spectrum, and the emission falls to zero at very long and very short wavelengths (Figure 17).

Planck found that he could bring theory into line with experiment if he introduced the idea that energy can be emitted or absorbed by a body only in 'buckets' of fixed size. The amount of energy in each bucket is given by the frequency of the radiation (which is related to the wavelength, as the velocity of the wave is equal to the product of frequency and wavelength) multiplied by a constant, h, now known as Planck's constant. Thus, the higher the frequency (or shorter the wavelength), the larger the bucket of energy must be. Planck called h the 'elementary quantum of action', and each 'bucket' is more properly referred to as a quantum of energy.

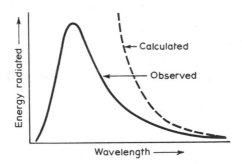

Figure 17 *The problem Planck set out to solve concerned the shape of the observed energy spectrum of radiation from a hot body. The calculation based on 'classical' physics agrees with observations at long wavelengths, but it implies that an infinite amount of energy is radiated at short wavelengths. In fact, the radiation peaks at a wavelength related to the temperature of the body; the higher the temperature, the shorter – more blue – the wavelength of maximum intensity*

Planck's basic discovery was that radiation from a glowing lump of coal, or any other body above the absolute zero of temperature, emerges not continuously like a stream of water from a tap, but more like the hail of bullets from a machine gun. We are not aware in the everyday world of this structure within radiation, for Planck's constant is so small that the individual buckets, or quanta, effectively constitute a smooth stream, in the way that minute dots make up the smooth blocks of letters on the printed page. The value of h is 6.626×10^{-34} joule seconds, so a quantum of visible light, of frequency about 10^{15} Hz, has an energy of $6.626 \times 10^{-34} \times 10^{15} = 6.626 \times 10^{-19}$ joules. This is roughly one thousand-millionth of the amount of solar energy reaching each square millimetre of earth each second.

It required the genius of Albert Einstein to be bold enough to contemplate the full consequences of Planck's discovery: to propose that not only did the radiating body lose its energy in quantized portions, but that the radiation itself travelled across space still packaged in the tiny bundles of energy. As Planck had done before him, Einstein took this new step forward in order to explain measurements that could be accounted for in no other way.

Experiments had shown that when ultraviolet light falls on the surface of certain metals, they emit electrons – this is the so-called photoelectric effect. But the results of the experiments were inconsistent with what one might expect were the electrons being knocked out by incoming *waves* of light. Einstein's masterstroke was to discover that he could reproduce the results exactly if he described the light not in terms of waves but in terms of *quanta* of energy in accordance with Planck's equation. Light, concluded Einstein, was behaving in this instance not so much as a continuous wave, but as a stream of 'wave packets', each carrying a quantum of energy. The comfortable notion that radiation travels through space as ripples in the electromagnetic field seemed not quite as secure as it had done at the end of the nineteenth century.

At about the same time, in 1905, Einstein also produced the work for which he is undoubtedly better known: his special theory of relativity. This too unsettled comfortable old ideas about the propagation of electromagnetic waves through space. Michelson and Morley had found that the velocity of light is independent of the earth's motion through space, but Einstein went farther than this and boldly postulated that the velocity of light is in fact a constant, the same to all observers in uniform motion relative to each other, irrespective of their velocity, and independent of the velocity of the source.

The theory of special relativity deals not just with light, but gives the correct equations for transforming variables, such as distances, times, velocities, and so on, from one frame of reference moving uniformly – at constant velocity – relative to another. The theory ensures that the laws of mechanics and Maxwell's equations of electromagnetism remain the same in any such reference frame, but in so doing it forces upon us some novel notions. The most famous of these, probably, arises from the fact that the equation for the energy of a moving body acquires an extra term, additional to the one which correlates with kinetic energy. This term is equal to mc^2, where m is the mass of the body and c is the velocity of light. It has become known as the rest-mass-energy, and showed for the first time the equivalence between mass and energy.

Physicists in the latter part of the nineteenth century had visualized electromagnetic energy travelling through space like the ripples on a pond. In the space of a few years at the turn of the century Planck and Einstein had upset the applecart. They showed that electromagnetic radiation often behaves more like a succession of wavelets than like a continuous rolling tide. The nature of the electromagnetic force began to emerge as something more subtle than Maxwell, Faraday and their contemporaries could ever have imagined.

In the following years more physicists grew to appreciate the importance of quantization, in particular once Niels Bohr had taken Planck's concept and used it to improve on Rutherford's model of the atom, as chapter 3 describes. And in 1923 the quantum nature of light took on a new meaning when Arthur Compton proved that the packets of light behave like particles – or 'photons'. The quantum of light has no rest-

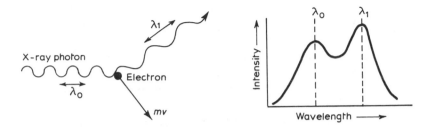

Figure 18 *Compton found that the wavelength of X-rays could increase when they were scattered by a thin metallic foil. His explanation was that the X-ray photons lose momentum in collisions with atomic electrons, just as if the photons were particles. He related the momentum of a photon to the inverse of its wavelength; photons of lower momentum have longer wavelengths*

mass-energy, only kinetic energy, but Compton's experiments showed that the wave packets could lose momentum and energy in collisions with electrons, rather as in a game of billiards. For a particle such as an electron, momentum is the product of mass and velocity, but clearly this definition has no meaning for the quantum of light. Instead, Compton followed earlier work of Einstein and assigned the photon a momentum of h/λ, where h is Planck's constant and λ is the wavelength. In scattering from an electron, a photon can lose momentum and shift to a longer wavelength, as Compton's experiments proved (Figure 18).

Prince Louis de Broglie took the next logical – albeit totally unprecedented – step in 1924: if electromagnetic radiation could behave as a stream of particles, he asked, why not treat particles as wave packets and assign them a wavelength, given by the inverse relationship, so that the wavelength is equal to h divided by the particle's momentum? Thus was born the theory of quantum mechanics, which when developed by Erwin Schrödinger and Werner Heisenberg became the second great theory to emerge in the twentieth century, to stand alongside special relativity.

Quantum mechanics provided a sound formalism that replaced Bohr's model of the atom and its various inadequacies. The theory treats a particle as a wave in the sense that the wave packet describes the probability of locating the particle in a given quantum state; the observed states – of electrons in an atom, for instance – correspond to solutions of a 'wave equation'. But there is more to the wave description of particles than mathematical trickery. Experiments show that particles in some instances do scatter from objects like waves do; the conceptually rational image of subatomic billiard balls does not stand up to close rigorous scrutiny, and our attempts to draw analogies between the everyday world and the atomic world begin to fall apart.

Quantum mechanics and relativity together form the basis of 'modern' physics, and they force us to modify our ideas about 'particles', 'forces' and 'fields'. Indeed, we find that all three concepts are intrinsically interwoven and tend to merge into each other. The first indications of change come when we write down a quantum wave equation to describe an electron with the correct relationship between energy and momentum that special relativity prescribes – in other words, an equation for the electron that is universally valid. This is precisely what Paul Dirac did in 1928 in his relativistic theory of the electron. And, in the solutions to his equation, Dirac made two vital discoveries. First, the electron must have an intrinsic angular momentum, or spin, with two possible values, $s = -1/2$ or $+1/2$, corresponding to two spin orientations, as described in chapter 3. Secondly, solving the relativistic electron equation involves taking square roots, and so it provided two sets of solutions, one of positive energy states, the other corresponding to states of negative energy. The electrons we observe in atoms, for instance, have positive energy; so what is the meaning of the states with negative energy?

Dirac's explanation was that under normal circumstances all the negative energy states are in fact fully occupied. The so-called exclusion principle, due to Wolfgang Pauli, states that no two particles of half-integral spin can share the same quantum numbers, or in other words occupy the same energy state. Once the negative energy states are completely occupied, one electron at a time, further electrons must occupy the positive energy states. The electrons in the world we observe are these positive energy particles; we are unaware of the 'sea' of occupied negative energy states. However, Dirac postulated, if a negative energy state becomes unoccupied, it will be equivalent to there being created a particle of opposite charge (positive) in a positive energy state, which we can therefore observe. Thus, if an electron is knocked out of the underlying sea of negative energy (unobservable) states into a positive energy (observable) state, a 'hole' will appear in the negative energy sea, or, according to Dirac, a positively charged, positive energy particle of the same mass as the electron will materialize along with the electron (Figure 19). In this way Dirac predicted the existence of the antielectron, or positron, discovered in 1932 in the tracks of cosmic rays through a cloud chamber.

The vindication of Dirac's theory was of vital importance, for in the way that Einstein's work on the photoelectric effect upset cherished notions about the nature of light, so Dirac's theory disturbs our concept of a particle, like the electron, as a simple building block of matter. The relativistic electron theory forces us to accept the creation of new matter in the formation of an electron–positron pair by the process of injecting sufficient energy into the unseen negative energy sea; the energy has to

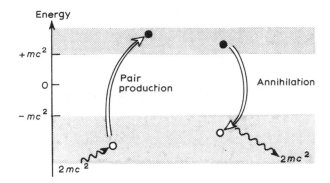

Figure 19 *The solutions to Dirac's relativistic electron equation imply the existence of negative energy states that are normally fully occupied. But an electron in one of these states can be knocked into a state of positive energy if it is given sufficient energy – at least 2mc². The hole left in the negative energy states corresponds to the formation of a particle with positive charge and the same mass m as the electron – the positron. In this way, an electron–positron pair can be created. Alternatively, an electron can fall back into a hole, releasing an energy of 2mc²; in this case the electron and positron annihilate each other and both disappear, leaving only a quantum of energy – a photon*

be at least $2mc^2$, where m is the mass of the electron and c the velocity of light. Equally, the electron and positron can 'annihilate', the electron falling back to the negative energy state, releasing a total energy of $2mc^2$. Thus is Einstein's equivalence of mass and energy manifest in Dirac's theory, proving matter to be more ephemeral than we at first supposed.

Dirac's theory not only upsets ideas about the permanence of matter but it immediately forces us to change our notion of empty space. We do not see the electrons with negative energy that are normally hidden, but once we spark that empty space with enough energy ($2mc^2$), we force particles – electrons and positrons – to materialize as if from nothing. Even a vacuum seems as if it is teeming with particle life waiting to be plucked out at the appropriate moment. This realization must, of course, influence our notions about forces and fields that operate over what we normally call 'free' space.

An exceedingly good theory

The connection between Maxwell's fields, Einstein's photons and Dirac's electrons and positrons is made in quantum field theory, developed by a number of theorists in the late 1920s. This began with attempts to introduce quantization into yet another realm of physics: to quantize the electromagnetic field. One result of this step into the unknown was a better understanding of the dual nature of light – as photons and as electromagnetic waves.

According to quantum theory, the vibrations of the electromagnetic field can have only fixed amounts of energy, given by Planck's relation. So although the field can vibrate at any frequency, the energy associated with that vibration cannot take any value, but is quantized. Moreover, we can identify the quantized vibrations with the photons that Einstein required to explain the photoelectric effect, and which existed implicitly in Planck's quantization of emitted and absorbed energy. Dirac consolidated this view of the the true nature of the photon when in 1927 he treated the electromagnetic field as a gas of photons and was able to calculate laws that Einstein had earlier shown to rule the emission and absorption of light by atoms. But in this work Dirac quantized only the radiation of the electromagnetic field – the photons. There was still the matter of the interaction between electrically charged particles and the electromagnetic field, as in the simple electrostatic interaction that Coulomb had found, which governs the behaviour of the atom.

Just as the dual nature of light had led de Broglie to associate 'matter waves' with particles such as electrons, so the identification of the photon as the quantum of the electromagnetic field led theorists to consider the electron as a field quantum. From there the obvious step was to attempt a theory that treated such 'particle fields' together with the electromagnetic field, so as to deal with the interactions of electrically charged particles and photons within the one theory. The result is the theory known as quantum electrodynamics, or QED for short, and it has turned out to be an exceedingly good theory, capable of making precise predictions of minute effects. But it took twenty years to evolve from the early work at the end of the 1920s.

In quantum electrodynamics, Maxwell's fields take life in the form of the invisible quanta – the photons – which flit from one charged particle to another. The photons carry momentum and energy between charged particles – or rather, the 'matter' fields – and so produce the effects that we observe as the electromagnetic force (Figure 20). If the particles carry the same charge, they repel each other; if they are unlike charged, they are mutually attracted. So now, in quantum electrodynamics, theory provides an answer for the mysterious action at a distance: interactions occur via the influence of a field manifest by its quanta which in effect carry the force between interacting bodies.

There is however one catch with this picture. In exchanging a quantum, the two particles (or rather, their fields) conserve energy and momentum overall, but at the instant that one particle 'releases' a quantum, and at the point at which the other 'catches' it, energy is not conserved. In

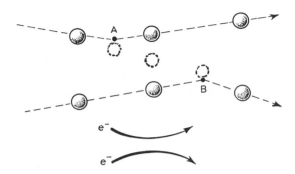

Figure 20 *We can visualize the interaction of two electrons in terms of the exchange of a field quantum – a photon – emitted by one electron at A and absorbed by the other at B. The photon carries energy and momentum, and in this example brings about the effect of a repulsive force. The emission and absorption processes do not conserve energy, so the exchanged photon is dubbed a 'virtual' particle; it is merely borrowed from the underlying field for the duration of the interaction*

classical physics this would strike an immediate death blow to the theory, but not so in quantum physics.

Werner Heisenberg discovered in 1926 that there is a fundamental limit to how precisely we can know all the details of a particular quantum system. We can, for example, know exactly the position of an electron, but if that is the case, we lose all information about its momentum. Conversely, if we know its momentum, we are uncertain as to its position. This 'uncertainty principle' can be expressed by stating that the product of the uncertainty in momentum and the uncertainty in position must be less than $h/2\pi$, where h is Planck's constant. This relationship is integrally tied up with the concept of an electron as a wave packet, which gives the probability that the electron is in a certain quantum state, but not a precise yes or no. An alternative form of Heisenberg's principle concerns the measurement of time and energy, and again the uncertainty in one times the uncertainty in the other must remain less than $h/2\pi$.

It is this aspect of quantum theory that comes to the rescue in the apparent failure of conservation of energy in quantum-field theory. According to the uncertainty principle, the energy of the quantum can be borrowed for a limited amount of time within the uncertainty $h/2\pi$; but the more energy borrowed, the shorter the time for which it can be borrowed. Thus the exchanged quanta are in a sense 'on loan' from the field, and must be returned once the time is up. For this reason they are called 'virtual' photons, to distinguish them from the 'real' photons that are the quanta of an electromagnetic wave (such as light propagating through the field).

The concept of a virtual particle goes beyond the photon, for in the same way the quanta of the particle fields – the electrons and positrons – can appear, briefly violating conservation laws, as virtual shadows of their normal selves. In some interactions we can imagine a virtual electron being exchanged, or a virtual photon creating a virtual electron–positron pair, which can then annihilate, providing the uncertainty relationship is not violated. This implies that the vacuum, far from being empty, is an extremely busy place, at least on a timescale short enough for Heisenberg's principle to allow the creation of particle–antiparticle pairs. If we could freeze this structure we would see a multitude of particles and antiparticles and photons; but, of course, we cannot, and this is exactly what the uncertainty principle means.

Why did it take twenty years for quantum electrodynamics to become fully developed? The problem with the earlier versions of the theory was that it could not cope with certain interactions, such as those that involve the emission and reabsorption of a photon by an electron. In such cases, the theory said that the probability for the interaction is infinite. A particular instance of this breed of difficulty concerns the existence of a single electron. The charged particle is the source of an electromagnetic field which QED allows us to visualize as a haze of virtual photons appearing and disappearing within the time allowed by the uncertainty principle; unfortunately we can conceive of this time as being reduced to zero, in which case the energy involved becomes infinite!

The solution to these troublesome infinities finally came about in the late 1940s, with work by Richard Feynman, Julian Schwinger, Sin-itiro Tomonaga and Freeman Dyson. The technique that these theorists developed independently is known as renormalization, and it allows us to cancel out most of the positive infinities with corresponding negative infinities. Those remaining can be swept away by defining the charge and mass as being the finite values that we measure. We can never measure the mass an electron would have without its cloud of virtual photons, for there is no such a thing as an electron without an electric charge. The infinite contributions to the theory from the infinite number of virtual photons thus become subsumed in the real, measured mass of the electron.

In addition to his work on renormalization, Feynman also made a great contribution to quantum field theory by inventing diagrams to aid calculations. These diagrams also help us to visualize the exchange processes, by representing time and space along two perpendicular axes; a straight line represents the life of a charged particle such as an electron, while a wiggly line depicts a photon (Figure 21). In this way we can depict, for example, the emission of a photon by one electron and its absorption by another; the creation from a photon of an electron–positron

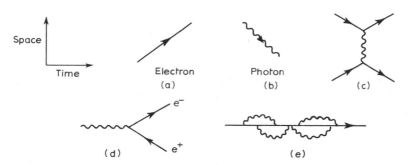

Figure 21 *Feynman diagrams are an important tool to aid calculations, but they also help us to visualize the processes of quantum electrodynamics. An electron is represented by a straight line in space–time (a), a photon by a wiggly line (b). Two electrons can interact by the emission and absorption of a virtual photon (c). A positron is equivalent to an electron moving backwards in time, so (d) represents pair production. A stationary electron appears as a horizontal line (e), with loops representing the continual emission and absorption of virtual photons*

pair; and the photon 'loops' that represent the kind of process that leads to the difficulties with infinities. A stationary electron with its cloud of virtual photons appears as a single straight line, moving only in the direction of time, but emitting and reabsorbing photons as time passes.

Feynman's diagrams also provide an alternative explanation for the negative energy states that Dirac discovered in his relativistic electron theory. According to Feynman, a solution corresponding to a negative energy particle moving backwards in time is equivalent to a solution for a positive energy *anti*particle moving forwards in time. The words 'backwards in time' do not imply anything about time travel, rather they refer to changing the signs of variables describing the particle state.

The field inside a proton

Quantum electrodynamics is still the best theory physicists have to describe a particular class of phenomena, despite the 'trickery' of renormalization that leaves a number of theorists, including Dirac, unsatisfied. However, even in the early days before renormalization had been invented to solve the problems, QED in its incomplete form still enabled theorists to calculate electromagnetic processes occurring in cosmic-ray interactions, such as the radiation of a photon by electrons passing close to atomic nuclei (a process known as *bremsstrahlung*, from the German for 'breaking radiation', as the electrons cast off the photons while being decelerated by the electromagnetic field of the nucleus). And so QED provided a model for theories of other interactions, such as those due to the force that holds the protons and neutrons within the nucleus, and those that underlie the radioactive decay of certain types of nuclei, and indeed of neutrons that have been liberated from a nucleus.

The electromagnetic force – as manifest in the electrostatic 'Coulomb' interaction between charges – is strong enough to hold electrons round the nuclei of atoms. A measure of the strength of the force is given by the so-called fine structure constant, denoted by α. α is a pure number, with no dimensions, and depends only on quantities that appear to be fundamental constants of nature: the electron's charge, the velocity of light and Planck's constant. Its value is approximately equal to 1/137. The name 'fine structure constant' derives from its appearance as a 'correction term' in expressions for the energy of an electron, which accounts for certain fine structure observed in atomic spectra. In quantum electrodynamics, α reappears as a 'coupling constant', in effect a measure of the probability for the emission of a photon.

According to Coulomb's law, the range of the electrostatic force is infinite – it decreases as the square of the distance from the source, so that infinitely far away its effect is infinitely small (that is, zero) and in practice it is too small to measure at large distances. In terms of the exchange of virtual photons, this infinite range reflects the fact that the quanta of the field have no rest-mass. It is possible in theory for the virtual photon to travel infinitely far without violating the uncertainty principle, because, being massless, it can have an infinitely small energy. What happens, however, in the case of a field quantum that does have mass? Its rest-energy is mc^2, by Einstein's relation, so according to the uncertainty principle, the time it can live must be less than $h/2\pi mc^2$. Even if the quantum is travelling very close to the speed of light, c, the maximum distance it can cover in this time is not infinite, but is $h/2\pi mc$. In other words, a force that operates via the exchange of a massive quantum has a limited range.

Such was the reasoning that led the Japanese physicist Hideki Yukawa to postulate a heavy field quantum for the nuclear force, which appears to have a limited range, in fact, the size of the nucleus – roughly 10^{-15} m. A particle travelling close to the speed of light will take $10^{-15}/c$ seconds, or roughly 10^{-23} s, to cross the nucleus. The maximum mass-energy it can borrow during this time, according to the uncertainty principle, is approximately $h/2\pi 10^{-23} = 10^{-34}/10^{-23} = 10^{-11}$ joules. This is about 10 per cent of the mass-energy of the proton.

Arguing along similar lines, Yukawa proposed in 1935 that the strong nuclear force is carried by virtual field quanta with a mass about one tenth that of the proton. Two years later particles of the appropriate mass were discovered in experiments with cosmic rays, but it turned out that these 'mesotrons' were not the particles of Yukawa's theory. Yukawa's field quanta, in real rather than virtual form, were not revealed until 1947, when improved experiments could show that there exist slightly heavier particles that decay rapidly into the 'mesotrons' discovered earlier. We now refer to the particles first observed as muons, while Yukawa's particle, which is slightly heavier, has become known as the pi-meson or pion.

Yukawa's theory enables us to picture the nucleus as a cluster of protons and neutrons immersed in a cloud of virtual pions which flit between the nucleons, binding them together in a tight cocoon of the strong nuclear field. The pion exists in electrically charged as well as neutral versions, and so can implement proton–neutron, as well as proton–proton and neutron–neutron interactions.

The concept of pion exchange can still be used to explain many aspects of the interaction between protons and neutrons, but it seems to be not as fundamental a process as the exchange of photons. Chapter 3 describes how the pion, like the nucleons and other particles, appears to be built from more basic entities, the quarks. We now believe that the true nature of the strong force lies in what happens between the quarks that make up the pions and the other particles.

The theory that has evolved during recent years to explain the force between quarks rests to some degree on analogy with quantum electrodynamics. It is known as quantum chromodynamics, where 'chromo' refers to 'colour', the rather confusing name for the property of quarks that is to quantum chromodynamics what electric charge is to QED.

The idea of colour first appeared in attempts to solve a paradox that arose in the quark model of baryons. The quarks must have spin $1/2$ if we are to explain the spin of baryons, and they therefore belong to a class of objects known as 'fermions'. Fermions are all those objects – whether fundamental particles such as quarks and electrons, composite particles such as baryons, or atoms such as helium-3 – which have a total spin of half-integral units, that is, $s = 1/2, 3/2$ . . . and so on. As such they are distinct from 'bosons', or objects with integral values of spin, $s = 0$, 1, 2, 3. . . . These include photons, mesons and atoms such as helium-4. In quantum theory a collection of fermions behaves differently from a host of bosons, for whereas many bosons can take up the same energy state, no more than one fermion at a time can occupy the same state, as the Pauli exclusion principle states.

Quarks are fermions, therefore they must obey Pauli's exclusion principle. But what happens, then, in particles such as the Ω^-, which comprises three strange quarks, all apparently spinning in the same direc-

Figure 22   *The Ω⁻ particle seems to contain three identical quarks, all spinning in the same way (a), a condition forbidden by Pauli's exclusion principle which does not allow quarks to be in identical quantum states. The solution to this paradox is to endow the quarks with another property, 'colour', which distinguishes them (b) and allows them to spin the same way within the Ω⁻*

tion (Figure 22), that is, all with the same spin quantum numbers? The solution proposed in 1965 by Oscar Greenberg of the University of Maryland was to endow the quarks with a new quantum number, dubbed colour, which comes in three varieties, red, green and blue, like the three primary colours of light. If the quarks in the Ω^- all have different colours, then the problem is solved (Figure 22); colour distinguishes them, and the quarks can have the same spin without upsetting Pauli's rule.

As happens so many times in the development of scientific understanding, what in one decade may appear as clutching at straws later proves to be of vital importance. The colour property of quarks now appears as something far more fundamental than a 'flag' to differentiate quarks. Indeed, it is as fundamental as the concept of electric charge, for the colour 'charge' on a quark acts as the source of a 'colour' field, in fact, the very field that manifests itself as the strong force.

Quantum chromodynamics is the quantum-field theory of the colour fields that emanate from the quarks, and the quanta of the fields are known as 'gluons'. These gluons are massless, like the photon, but differ from the photon in one crucial way. Whereas the photon carries no electric charge and is therefore itself impervious to the electromagnetic force that binds electrons with nuclei, the gluons do carry colour charges and so they themselves generate colour fields, as do the quarks. There are eight varieties of gluon, each with different combinations of colour and anticolour (as carried by the antiquarks).

The nature of the colour force seems very different from the electromagnetic force, and this is due in part to the fact that the gluons can themselves interact with the colour fields. Indeed, the force seems such that single quarks can never leave a cluster, for no single quark has ever been observed. This is like saying that colour charges cannot exist alone, as electric charges can, but must form only neutral clusters, rather as north and south magnetic poles must always appear together in a magnet. Thus the three colours of the quarks in a baryon must always neutralize each other, as must the colour and anticolour of the quark–antiquark pair of a meson. The colour force seems such that we can observe only 'white' particles in nature. Nor does the colour force become weaker with distance as does the electromagnetic force. On the contrary, it seems to grow very strong at distances comparable to nuclear dimensions – 10^{-15} m – and although the gluons are massless, the range of the force is restricted and is by no means infinite. But at much smaller distances the colour force appears relatively weak, so that the quarks act as if only loosely bound to each other when we observe them in the appropriate circumstances.

The discovery of colour charges and colour fields does not invalidate Yukawa's picture. We still visualize the protons and neutrons of a nucleus as bound by virtual pions – rather as, in some molecules, electrostatic forces bind collections of electrically neutral atoms. Indeed, in high-energy collisions between protons and neutrons, such as occur at particle accelerators or in the interactions of cosmic rays, we can observe mesons – kaons as well as pions – 'boiling off' from the energized nucleons, as quark–antiquark pairs are created from the colour fields.

The electromagnetic force and the strong nuclear force can thus be described in terms of quantum-field theory, although admittedly our understanding of QCD still falls a long way short of the exactness of QED. But what of the weak nuclear force, which underlies the decay of the neutron, the pion and many other particles? Attempts to develop a similar theory for this force have proved so pivotal in our progress towards a 'unified' picture of all the forces that this story deserves a chapter to itself.

5

The Force that Drives the Sun

The sun has shone, according to geologists, for 4.5 thousand million years or so. It shines, according to astrophysicists, by converting hydrogen, the lightest element, into helium, the second lightest, and releasing energy as it does so. So far, the sun has converted in the region of 10^{26} tonnes of hydrogen into helium, and it has sufficient fuel left to carry on shining the same way for at least another five thousand million years. The energy in the light and heat we receive on earth has been through many processes before it arrives here, and bears no memory of the nuclear reactions in which it originated; indeed, it takes over ten thousand years for a photon to fight its way out, being absorbed and re-emitted many times in the variety of layers of gas and plasma that form the sun. However, there are other messengers from the sun that bring with them the imprint of their birth. These are the solar neutrinos, produced in the very interactions that spark the solar burning and initiate the conversion of hydrogen to helium. Although the sun's energy that is so vital to life on earth comes in electromagnetic form, as light and heat, the neutrinos betray the nature of these important initial reactions. Of nature's four forces, the only force neutrinos feel is the weak nuclear force, and it is the weak force that fuels our sun.

The conversion of hydrogen to helium in the sun's core involves the fusion of four protons – hydrogen nuclei – to form a helium nucleus (^{4}He), which comprises two protons and two neutrons (Figure 23). To conserve electric charge, two positrons are emitted in the process; to conserve the total lepton number, two neutrinos (of electron type) are produced. In addition, two photons (γ) emerge, so we can write down the overall reaction as:

$$4p \rightarrow {}^4He + 2e^+ + 2\nu_e + 2\gamma + E.$$

The reaction releases a net energy, E, of 24.3 MeV, or 4×10^{-12} joules, which provides a total of 4×10^{14} joules for every kilogramme of hydrogen 'burned' – and roughly 1000 million tonnes burn each second.

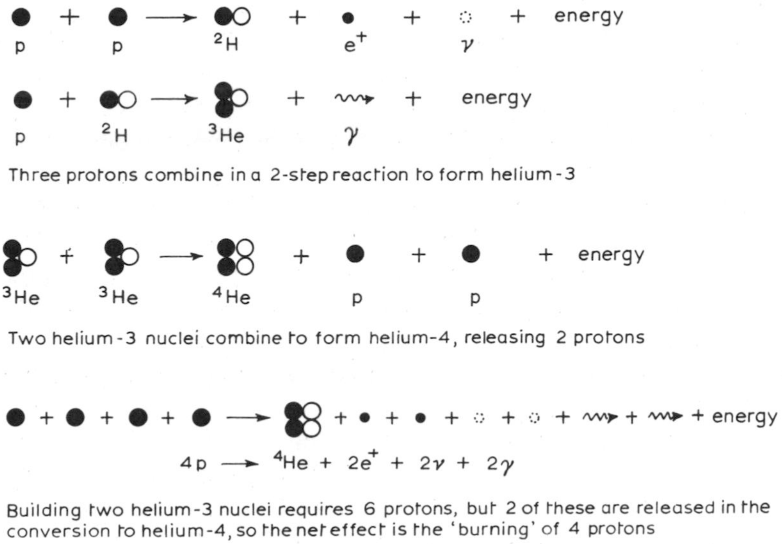

Three protons combine in a 2-step reaction to form helium-3

Two helium-3 nuclei combine to form helium-4, releasing 2 protons

$$4p \longrightarrow {}^4He + 2e^+ + 2\nu + 2\gamma$$

Building two helium-3 nuclei requires 6 protons, but 2 of these are released in the conversion to helium-4, so the net effect is the 'burning' of 4 protons

Figure 23 *The basic proton–proton chain that fuels the sun, with the release of energy every time four protons form a helium-4 nucleus; the first two steps must happen twice for every occurrence of the third step. Other variations on this theme also occur, but less frequently*

The conversion takes place via a number of intermediate interactions, known together as the proton–proton chain, because the whole process begins with a reaction between two protons to form the nucleus of deuterium or 'heavy hydrogen', which consists of a proton and a neutron. Deuterium (^{2}H) is an isotope of hydrogen; that is to say, it has the same number of protons (one) and therefore the same number of electrons (one), and therefore the same chemical properties. The only difference is in the extra neutron, which makes deuterium heavier. In forming a deuterium nucleus, one of the original protons converts into a neutron, emitting a positron and a neutrino:

$$p + p \rightarrow {}^2H + e^+ + \nu_e.$$

The weak interaction, which is responsible for this process, is so feeble that even in the sun's core, where the particles are 100 times more densely packed than in water, a hydrogen proton will survive some ten thousand million years before succumbing to the pp reaction that starts the process of hydrogen 'burning'. Clearly if the weak force were not so weak, our universe, and our solar system in particular, would be a very different place.

A first theory

What is this force that fuels our sun? Although man has observed the sun for thousands of years, our understanding of the precise reactions within the solar retort is by no means perfect. Clues to the processes deep within the sun come indirectly from observations of the energy it emits and from its composition, which can be determined from detailed studies of the sun's spectrum. A laboratory simulation of the solar interior is precluded by the long time it takes for the initial pp reaction. However, the weak force mediates other faster reactions which can be studied in the laboratory, and which provide a test bed for our ideas about the weak force, if not for the astrophysics of the sun's burning.

In the closing years of the nineteenth century Ernest Rutherford, working at McGill University in Canada, discovered that uranium emits two kinds of radiation, one that a few micrometres of aluminium will easily stop, and another that can penetrate a few millimetres of the same material. Rutherford called the weakly penetrating radiation alpha rays, the other beta radiation. Moreover, he discovered that the alpha rays carried positive electric charge, while the beta rays were negative. Subsequently, some ten years later, in 1909, Rutherford proved that the alpha rays were the nuclei of helium atoms, in other words, conglomerations of two protons and two neutrons. However, in 1900, only a year after Rutherford had discovered them, Henri Becquerel, working in Paris, was able to show that the beta rays were identical to cathode rays – in other words, that they were streams of energetic electrons.

We now know that the electrons emitted by uranium, and many other kinds of nucleus, are not explicitly contained within the nucleus; rather, they materialize when a neutron converts into a proton, in the process that has become known as beta decay:

$$n \rightarrow p + e^- + \bar{\nu}_e.$$

The emission of the neutrino (actually an antineutrino as the bar above indicates, thus conserving the lepton number, as described in chapter 3) was predicted in 1930 by Wolfgang Pauli, although no such particle was known to exist at the time. Experiments showed that the electrons emitted in beta decays carried a range of energies, which would not be the case were only two particles, the proton and electron, the results of the decay. This observation led some physicists, including Niels Bohr, to speculate that energy and momentum might not be conserved within the weird domain of the atomic nucleus. Pauli came to the rescue, however. Rather than abandon conservation of energy and momentum, he preferred to invent a new particle – a rather daring move that he was cautious in springing on an unsuspecting world. The new particle had to be electrically neutral, and virtually massless, as in some cases the electrons could carry away more or less the maximum amount of energy expected. By 1934 Enrico Fermi, at the University of Rome, had accepted Pauli's hypothesis and named the particle the 'neutrino' (for 'little neutral one'), and had incorporated it into his new theory of beta decay.

Fermi's theory is the first real theoretical model for the action of the weak force, which is responsible for beta decay. As a theory it is incomplete; it does not reveal why the weak interaction is weak or what the nature of the weak force is, nor does it work at high energies, as we shall see. But the theory works remarkably well at low energies and allowed physicists in the 1930s to calculate the reaction rates for beta decays and the shape of the spectrum of energies of the emitted electrons.

Fermi looked to quantum electrodynamics (QED) for his inspiration, which at the time Dirac and others were showing to be so successful. By analogy with the way an electron radiates a photon, Fermi proposed that a neutron radiates simultaneously an electron and an antineutrino, so changing into a proton (Figure 24a). On a Feynman diagram, the electron–antineutrino pair is emitted at one point in space–time, just as a photon is. Moreover, by further analogy to QED, Fermi conceived of beta decay as the interaction of two 'currents' (Figure 24b), one carrying the heavy particles (the neutron or proton) and one carrying the light particles (the antineutrino or electron). This step becomes more obvious if we rewrite the basic beta decay process as:

$$\nu_e + n \rightarrow p + e^-.$$

Note that in moving the antineutrino to the left-hand side it has become a neutrino, because in rewriting the reaction I have essentially reversed the direction in space–time of this particle. Strictly speaking, the two currents are the 'hadron current' and the 'lepton current', the proton and the

Figure 24 *In describing the beta decay of a neutron (a), Fermi assumed that the neutron (n) 'radiates' an electron (e) and an antineutrino ($\bar{\nu}$) as it converts into a proton (p). Taking the analogy with electromagnetism a step farther, Fermi conceived of the interaction of two currents (b), one carrying the hadrons (n and p) and one the leptons (e and $\bar{\nu}$). Note that the currents change charge in the beta decay process, which has been redrawn here as the interaction of a neutrino (v) and a neutron*

neutron being hadrons, while the electron and the neutrino are leptons.

Naturally, there are some important differences between Fermi's inter-action and QED. In the electromagnetic scattering of an electron and a proton, for example, the particles exchange a photon, while keeping their identity and electric charge:

$$e^- + p \rightarrow e^- + p.$$

In this case the hadron (proton) and the lepton (electron) currents conserve electric charge. In Fermi's theory of beta decay, however, the neutron changes to a proton, and the neutrino to an electron, so the two currents change electric charge at the point of interaction; the beta decay process is charge-changing (although, of course, charge is conserved overall). Moreover, in QED the radiated photon is the massless quantum of the electromagnetic field; the electron–antineutrino pair in Fermi's process is clearly something quite different, being neither massless, nor even a single particle.

However, Fermi was able to pursue his analogy between QED and beta decay far enough to work out the probability for a beta decay to occur. He found that it is proportional to the square of a constant, G_F, which can be calculated from the measured decay rates. G_F is related to Planck's constant, the velocity of light, and to the mass of the proton – all fundamental constants of nature. By contrast, in QED the probability for the emission (or absorption) of a photon is proportional to α, the 'fine structure constant', as we saw in chapter 4. Moreover, the probability for an interaction that involves both the emission and the absorption of a virtual photon, such as electron–proton scattering, is proportional to α^2, rather as the probability for beta decay is proportional to G_F^2. The constants α and G_F allow us to compare and contrast the weak and electromagnetic interactions in the following way.

There are two important differences between the two constants. First, G_F has dimensions – in other words, it is not a pure number but must be expressed in terms of units. In the case of G_F, the appropriate units are those of the inverse of mass squared, or $(mass)^{-2}$. The fine structure constant, α, on the other hand has no dimensions. This suggests again that Fermi's theory is not quite the same as QED. Secondly, G_F is numerically much smaller than α. In terms of proton masses, G_F is 'approximately equal to' 10^{-5}. This reflects the weakness of the weak interactions with respect to the electromagnetic force, for which the relevant constant $\alpha = 1/137$.

The beta decay of the neutron is only one of three related processes that Fermi's theory describes. All of these cause the transmutation of elements by changing the total number of protons, for it is the number of protons (and hence electrons) that determines the chemical nature of an atom. The decay of a neutron increases the number of protons by one, and thus transforms the element. Moreover, because the neutron is slightly heavier than the proton, by about 0.1 per cent, a free neutron decays eventually into a proton, an electron and an antineutrino, after an average lifetime of fifteen minutes. Within the confines of the nucleus, however, other forms of beta decay are possible. A proton can change into a neutron, emitting a positron and a neutrino. And a nucleus can capture a nearby electron, absorbing it in a process that converts a proton into a neutron, at the same time emitting a neutrino. These three kinds of beta decay all involve more or less the same particles – the proton, the neutron, the electron and the electron–neutrino (or their antiparticles) – so it is not surprising that the same constant, G_F, applies to each process. However, experiments in the 1940s and 1950s were to show that G_F is of more fundamental importance than it at first appeared to be.

In the 1940s, studies of the decays of the muon revealed that it could not be the particle that Hideki Yukawa had proposed as the mediator of the strong nuclear force binding protons and neutrons within the nucleus. Instead, the muon seemed to be a new electron-like particle, responding not to the strong force, but only to the electromagnetic force and the weak force, via which it decays. As with the beta decay of a neutron, we can visualize the muon decaying into an electron, a neutrino and an antineutrino (Figure 25), in a 'point interaction' involving four fermions – particles with spin $s = \frac{1}{2}$:

$$\mu^- \rightarrow e^- + \nu_\mu + \bar{\nu}_e.$$

This reaction involves only leptons – e, μ and ν – unlike neutron decay which involves the hadrons – p and n. So it comes as something of a surprise to learn that the strength of the reaction, which we can call G_μ, turns out to be almost identical to that for beta decay, G_F.

This observation underlies the concept of 'universality' of the weak force: the idea that the strength of the weak force is the same for all manner of particles, whether leptons or hadrons. Such an idea is not new. The universality of the electromagnetic force is borne out by the exact similarity of the electric charge of the electron and the proton; the force does not distinguish between leptons and hadrons, only between charged and uncharged particles. Similarly, the weak force does not distinguish

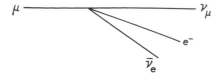

Figure 25 *The decay of a muon (μ) resembles the beta decay of a neutron and can be represented by a similar 'four-fermion interaction' at a single point in space–time*

between hadrons and leptons, nor even between charged and uncharged particles. Instead, it seems to depend on some universal 'weak charge', manifest in the coupling constant G_F.

The parity revolution

The step of proposing the universality of the weak force was taken by Richard Feynman and Murray Gell-Mann. In a paper published in 1958, the two theorists from Caltech put forward a new version of Fermi's theory to account for some recent results on the beta decay of cobalt, which had revealed a surprising asymmetry in the action of the weak force. The results were surprising in that the existence of such an asymmetry in nature was contrary to nearly everyone's intuition; they were not surprising in that they confirmed the work of two theorists published in 1956, a few months before the experiment in 1957.

Tsung Dao Lee, at Columbia University, and Chen Ning Yang, at the Institute for Advanced Study at Princeton, had been considering some problems concerning the decay of charged K mesons, or kaons, which had been discovered a few years previously in cosmic-ray experiments. An apparent paradox in the decays could be removed from the results by making one bold assumption – that weak interactions, such as the decays of K mesons, are not symmetric with respect to a mirror world, or more technically to inversion in space. Now this latter statement does not imply that measurements of weak interactions give different results if the experiment is turned upside down. Rather, it means that if the mathematical description – the quantum mechanical wave function – of the particles involved has the signs of all its spatial coordinates changed, then the new wave function describes something that we do not observe, nor, in fact, can we observe it. This changeover in coordinates is part and parcel of the concept of 'parity' in particle physics, which is worth considering in more detail.

In a Cartesian coordinate system where the three dimensions of space are represented by three axes, X, Y and Z, a point in space is denoted by its position along each of these three axes (x, y, z). The point where the three axes cross is the origin, where $x = 0$, $y = 0$, and $z = 0$, and because each axis can extend outwards in two opposite directions from the origin we can have both positive and negative values of x, y and z. Convention dictates that we use a 'right-hand rule' to assign the directions of the positive axes, with x, y and z as the thumb, forefinger and middle finger of the right hand (Figure 26a); the negative axes follow the directions given by the left hand. The so-called parity operator is in effect a 'magic wand' that changes the sign of all values of x, y and z. So a point

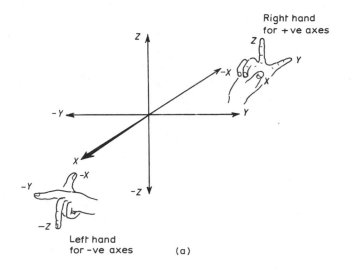

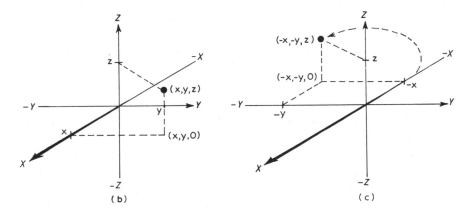

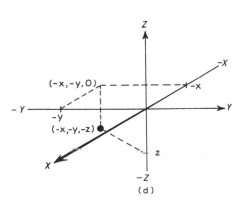

Figure 26 *In a right-handed coordinate frame (a), the* X, Y *and* Z *axes which mark out the three dimensions of space point in mutually perpendicular directions, like the thumb and first two fingers of the right hand. The negative axes point in the directions given by the left hand. The operation of parity takes a point from (+x, +y, +z) to (−x, −y, −z). It is equivalent to taking a point (b), rotating it through 180° about the* Z *axis (c), and then reflecting it in a mirror in the* XY *plane (d)*

in space at $(+x, +y, +z)$ becomes the point $(-x, -y, -z)$ and vice versa. This operation is in fact equivalent to rotating the point through 180° about the Z axis, and then reflecting it in an imaginary mirror lying in the plane defined by the X and Y axes (Figure 26b and c). Note that in a sense the point moves from a right-handed coordinate system to a left-handed system: we have interchanged what we mean by right and left.

If we imagine there to be an electric charge at the origin, then we know that the electromagnetic field at the point (x, y, z) is just the same as at the point $(-x, -y, -z)$, for it depends on the *square* of the distance from the charge, and this is the same for the two points. In other words, the field remains the same after the parity operation – we say that parity is conserved. Parity conservation is in fact a general property not only of all electromagnetic interactions but also for all reactions via the strong nuclear force. In these two cases, nature seems to show no preference for whether we choose to define a point as being at $(+x, +y, +z)$ or $(-x, -y, -z)$. Such a symmetry is intellectually appealing rather as the symmetry of a snowflake is aesthetically appealing. After all, our conventions as to what is positive, what is negative, which is the right hand and which the left are purely arbitrary. While we accept certain asymmetries in the world about us, such as the occurrence of the human heart towards the left side of the body, it still seems natural that, at a fundamental level, the forces that mould the world should make no distinction between left and right, or any other arbitrary convention. So, until 1956, no one ever queried whether parity symmetry holds for the weak decays of particles and nuclei.

At the level of subatomic particles, where the language of quantum mechanics must be used, the conservation of parity is manifest as the conservation of a quantum number, P, for parity. The value of P labels a particle according to its behaviour in the parity 'mirror'. If the 'wave function' – the quantum mechanical description – of a particle remains the same in the parity mirror, then its parity is said to be 'even', and $P = +1$; such a wave function is symmetric. A wave function that changes sign in the parity mirror is said to have 'odd' parity, and $P = -1$; such a wave function is antisymmetric. By defining the parity of the proton as even, so that $P = +1$, it is possible to calculate the parities for other particles and so evaluate the net parity before and after an interaction. Parity is conserved if this net value remains the same.

Lee and Yang observed that the weak decays of the K mesons could be explained if parity is not conserved in weak interactions. Moreover, they pointed out, there was no experimental evidence to disprove this claim, for physicists had the beauty of symmetry under the parity operation so firmly entrenched in their minds that no one had ever bothered to test whether it held in weak interactions. However, Lee and Yang suggested a number of experiments to check their ideas and so sparked off an immediate flurry of experimental activity. Within a few months, in 1957, the conclusive evidence that the weak force does not respect the conservation of parity came from an experiment on beta decay, and later that year Lee and Yang received their Nobel prizes.

In their decisive experiment, Chien Shung Wu, from Columbia University, and Ernest Ambler and collaborators, at the National Bureau of Standards in Washington, studied the beta decay of the nuclei of cobalt-60. The nuclei have an intrinsic spin and so behave like tiny magnets, which can be aligned with a magnetic field, provided the temperature is cold enough to slow down the normal thermal jiggling. By cooling their sample to 0.01 K – a hundredth of a degree above the absolute zero of temperature – Wu and Ambler were able to align the nuclei magnetically, and observe the electrons emitted as the cobalt-60 nuclei decayed to nickel-60.

If the weak force has no preferred direction in space, then the directions of the emitted electrons should reflect this symmetry. But Wu and Ambler found the opposite to be true: most of the electrons emerged in the same general direction with respect to the orientation of the nuclei – opposite to the direction in which the nuclear magnets were pointing. Moreover, with the direction of the magnetic field reversed so that the nuclei pointed the opposite way, the result still held: most of the electrons were emitted 'backwards' relative to the spinning nuclei.

The ramifications of these results become clearer if we consider what the experiment looks like in the mirror world that the parity operator provides access to. Suppose the magnetic field is in the direction of the Z axis. If a nucleus spins like a right-hand screw about the field direction, then the nuclear magnet is aligned with the field. In the real world it spins *clockwise* when viewed from the plane of the parity mirror – the XY plane (Figure 27a). The results of Wu and Ambler showed that more electrons were emitted in the direction from which the nuclei appear to be spinning clockwise, that is, towards the mirror plane in the diagram.

Now consider what happens in the parity-reflected world. In this case, the absolute directions of the electrons are reversed, although the nuclei continue to spin the same way (Figure 27b). This means that most of the electrons seem to be emitted in the direction from which the nuclei appear to be spinning *anticlockwise*, in contradiction to what happens in the real world. Thus the operation of parity turns the experiment into something that is not observed, something which does not exist in the real world – in other words, parity is not conserved.

If we turn to consider the electron and the antineutrino emitted by the nucleus, we come to some even more startling conclusions. In the decay, the intrinsic spin of the nucleus changes by one unit, so this amount of

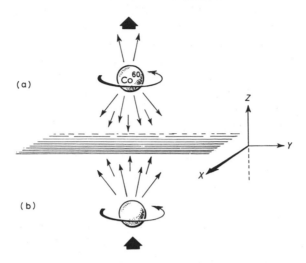

Figure 27   *Wu and colleagues aligned the nuclei of cobalt-60 in a magnetic field, so that the spins were nearly all in the same direction. In diagram (a), the Co⁶⁰ nucleus spins anti-clockwise as viewed from the XY plane, and the broad arrow indicates the spin direction given by the right-hand-screw rule. The experiment showed that more electrons were emitted in the direction opposite to the average spin direction – towards the XY plane in the diagram. In the parity-reflected world (b), the nucleus spins the same way, but the directions of the electrons are reversed, and we find a situation that is not observed in the real world. Thus we can say that parity is not conserved*

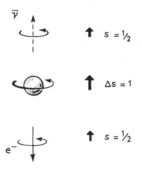

Figure 28   *In the beta decay of cobalt-60, the spin of the nucleus changes by one unit ($\Delta s = 1$), which must be carried away by the electron (e) and the antineutrino ($\bar{\nu}$). This implies that the spins of the two outgoing particles (small thick arrows) must be in the same direction. But the two particles leave the nucleus in opposite directions and this in turn implies that they have different spins relative to their directions of motion. Indeed, the antineutrino behaves like a right-hand screw, turning clockwise about its direction of motion; the electron is like a left-hand screw, and turns anticlockwise*

spin must be carried away by the two leptons. The electron and the antineutrino each have spin of half a unit, $s = \frac{1}{2}$. So to carry one unit between them, they must both spin the same way, in the same direction as the nucleus is spinning, to keep the total spin constant (Figure 28). Wu's experiment showed that the electrons are emitted preferentially backwards relative to the direction of spin of the nucleus. This in turn means that the electrons must be spinning anticlockwise about their directions of motion, like left-hand corkscrews. We say that these electrons are left-polarized.

Imagine watching one of these left-polarized electrons travelling away from us, and suppose that we can accelerate to overtake it. Looking back at it, the electron will now appear to us to be spinning clockwise as it recedes, like a right-hand screw thread. In other words, it will appear right-polarized! So an electron's polarization appears to change with our point of view, which makes it rather surprising that the electrons emitted by cobalt-60 should be polarized only one way. But what about the antineutrinos emitted at the same time?

The experiments with cobalt-60 did not reveal the antineutrino's direction, although with the benefit of hindsight I have added it to Figure 28; it required another experiment in the same year to study the polarization of neutrinos. Return for a moment to our imaginary experiment, but this time with a neutrino rather than an electron. The neutrino, being apparently massless, travels at the speed of light, which means that we can never overtake it and see its polarization change. It therefore seems feasible that neutrinos exist with only one polarization, and antineutrinos with the opposite polarization. Indeed, a clever experiment observing the reaction when a nucleus of europium-63 captures one of the inner atomic electrons and converts to samarium-63 showed that neutrinos are left-polarized – they spin anticlockwise about their direction of motion.

So, the antineutrinos emitted by cobalt-60 have their polarization already fixed by nature, and spin only one way: they are right-polarized. This in turn fixes the polarization of the electrons. The intrinsic screwhead of the neutrinos forces a difference between left-handed and right-handed in the physical world, and thus parity is violated in the weak interaction, the only interaction the neutrinos feel. There is, after all, an asymmetry in nature, an asymmetry that was for a long time totally unsuspected.

A second theory

By 1958, experiments had confirmed parity violation in the decays of muons and π- mesons; they had determined that neutrinos (antineutrinos) are naturally left-polarized (right-polarized); and they had studied the left-polarization of the electrons emitted in beta decay. We live, it seems, in a universe that is intrinsically left-handed, and the weak force betrays this handedness while the strong and electromagnetic forces remain oblivious to it.

Where did this bout of activity leave Fermi's theory? Fermi had not allowed for parity violation, and his theory incorporated only 'vector' currents in strict analogy with electromagnetism. These are currents that behave like a vector does under the parity operation. Now, the direction of a vector – a quantity like velocity in which direction is important as well as size – changes in the parity mirror, rather as the direction of the electrons from beta decay does. In other words, a vector quantity has odd parity. But Fermi's 'vector interaction' could not explain the decays like that of cobalt-60 where the nucleus changes spin by one unit. It turns out that the appropriate interaction in this case is one that involves so-called 'axial vectors'. The prime example of an axial vector is angular momentum, such as spin, the direction of which remains the same under the parity operation: it has even parity.

All these pieces of the puzzle – the violation of parity, the handedness of neutrinos, the vector and axial-vector nature of interactions – come together in the so-called 'vector-minus-axial-vector' or 'V–A' theory written down in 1958 by Feynman and Gell-Mann. And in the very same theory they included their concept of the 'conserved vector current', which by analogy with electromagnetism produces the same strength of weak interaction in nucleon decay as in muon decay. The theory proved a remarkable success; the two theorists were even bold enough to suggest that existing data which disagreed with their theory were wrong. As further experiments showed, Feynman and Gell-Mann were right.

The V–A theory provides a highly successful prescription for calculating processes at low energies – the decays of particles, such as neutrons, muons and pions (and with some important modifications, the strange particles), and interactions observed at accelerators prior to 1973. We must wait until chapter 7 to discover why that date is important, but long before then it was clear that, despite the precision with which V–A theory can be used, it does suffer from some obvious flaws. One fatal problem concerns the probability for an interaction such as that between a muon–neutrino and an electron to create a muon and an electron–neutrino:

$$\nu_\mu + e \rightarrow \nu_e + \mu.$$

At high energies the theory predicts a probability that is higher than it should be according to the simple idea that you cannot produce something from nothing. To overcome this difficulty, the original idea of Fermi's, of four particles interacting at a point via two currents, needs adapting in some way.

The weak 'photon'

It is worth recalling now that although Fermi wrote down his theory by analogy with QED there were some fundamental differences. One

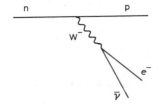

Figure 29 *It is possible to visualize beta decay via a photon-like carrier – an intermediate vector boson, W. The neutron radiates the W, which must be negatively charged, and converts to a proton; the W then decays into an electron and an antineutrino*

important difference was that Fermi's theory drew a comparison between the electron–antineutrino pair radiated in nuclear beta decay and the photon radiated by an electric charge. But the pair of leptons is clearly not the same breed of beast as the photon of QED. If there is to be a good analogy between the weak and electromagnetic interactions, then surely the weak force needs some carrier, some 'photon' of its own, to mediate the weak interactions (Figure 29).

We know already that such a carrier must be different from the photon; the range of the electromagnetic force is infinite, because the photon has no mass and can in theory be borrowed for as long a time as possible to give as long a range as possible, provided its energy is small enough. The weak force, on the other hand, is confined to dimensions the size of the nucleus. Furthermore, the success of the V–A theory, based as it is on the idea of a point contact between four particles, shows that at least at low energies the range of the weak force is so small that it appears to take place at a point. Only when we probe smaller areas of space–time in experiments at high energies can we begin to see that the point interaction is not the true picture. And given that the range of the weak force is very small, then we also know from the argument Yukawa put forward concerning the carrier of the strong nuclear force (chapter 4) that the particle exchanged – the weak 'photon' – must have some mass. Moreover, the weakness of the weak force, with reactions that occur in the range 10^{-10} to 10^3 seconds, as witnessed by the lifetimes of particles from the Λ hyperon to the neutron, suggests that the weak photon must be quite heavy – heavier than the mesons of Yukawa's theory, heavier indeed than the proton.

Yukawa was in fact one of the first theorists to consider the weak nuclear force in terms of an exchanged particle, when he attempted at first to explain both strong and weak nuclear interactions in a theory of mesons. Strong interactions, he conjectured, are between mesons and nucleons; weak interactions between mesons and leptons. However, Oscar Klein is now recognized as being the first to present the idea of an 'intermediate vector boson' – a particle like the photon with a spin of one unit. The label 'vector' refers to the fact that the spin is 1 and not 0, so that the particle's wave function behaves like a vector. Klein's hypothetical boson differed from the photon in that it was massive; his theory did not predict a mass, but it can be related to Fermi's constant, G_F, for the strength of beta decay, to give a value of 100 times the proton's mass. Furthermore, the new boson had to be electrically charged, to mediate the charge-changing reaction of beta decay. These have remained the outstanding characteristics of the proposed carrier of the weak force for nearly fifty years. The W particle – named for the weak interaction – is charged, massive and has spin 1.

As we shall see, it has taken the best part of fifty years to track down this elusive particle. For now, we must note that its introduction into Fermi's theory or into the more sophisticated V–A theory does not solve the problems. The 'intermediate vector boson model' still produces unrealistic probabilities, albeit at much higher energies. This problem is tied in with other difficulties with the theory which can be summarized in one bald statement: the theory is not renormalizable. The situation with the theory of the weak force in the 1950s echoed the difficulties with QED after its inception in the 1930s. As we saw in chapter 4, QED at first appeared to be riddled with infinities that plagued calculations of even some of the most fundamental processes, like the existence of a single electron. Renormalization was the cure for this disease – a prescription that allowed you to cancel out various recalcitrant contributions to a calculation. In the late 1950s the theory of the weak force was in need of such a cure. How this cure involved a deeper connection with the electromagnetic force than many people had dared imagine we shall see.

6

Towards a Theory

The *Concise Oxford Dictionary* lists five definitions for the word 'theory', which only goes to show how a word can have different meanings for different people. A theory can, for instance, be pure speculation, as in the statement 'I think the moon is made of green cheese'. A physicist would call this kind of theory a model and would proceed to draw from the model conclusions that he could test. (I can, for example, calculate the density of the moon using Newton's mechanics, and compare my answer with that measured for green cheese.) In this sense, Rutherford's concept of the atom as a miniature solar system was a model, based admittedly on some sound observations, which Bohr developed to perform calculations that could be verified by comparison with measured atomic spectra. But even Bohr's work fell short of providing what many physicists would regard as a 'proper' theory, in the sense given by the first definition in the *Concise Oxford Dictionary*: a theory is a 'supposition or system of ideas explaining something, *especially one based on general principles independent of the facts, phenomena, etc. to be explained*'.

The italics are mine, and I use them to emphasize the character of what I believe to be a good theory, at least as far as a physicist is concerned. A theory should *not* be an *ad hoc* set of equations that together happen to describe what we observe, and which provide what physicists refer to as phenomenology – 'the description and classification of phenomena'. It is fair to say that Bohr's atomic 'theory' was phenomenological, based on postulates for which there was no good evidence except that together they fitted the observations (not that this should detract from Bohr's achievement). It took Schrödinger, Heisenberg and others to provide something approaching a respectable *theory* of the atom within the framework of quantum mechanics, which provides a set of general principles, just as the *Concise Oxford Dictionary* demands. So, the word 'theory' can mean three things to a physicist – a model, a phenomenological set of rules, or a theory rooted in general principles forced upon us by the way that nature really is.

How do the theories of nature's forces we have encountered so far stand up to this semantic scrutiny? Quantum electrodynamics certainly appears a good theory, set soundly on the general principles of special relativity and quantum mechanics, which themselves seem to be manifestations of the nature of our universe rather than mere mathematical descriptions. Admittedly, it relies on the process of renormalization, which some physicists regard as *ad hoc*, but it goes a long way towards fulfilling our definition of a good physical theory. Notice here how we have turned the tables on the definition of a theory as pure speculation; to a physicist a theory is something he believes provides a good description of nature, not some fanciful idea that exists in the mind alone.

From QED flows the notion that fields are in some sense more fundamental than either particles or forces, and the properties of the forces and particles we observe arise from the nature of those fields.

However, QED not only leads us to concepts that are deeper than those with which we began; it has two further characteristics that are essential to a physical theory deserving of the name. It makes predictions that we can test, and it does not under any circumstances produce impossible answers. QED can be used to calculate minute changes in the energy levels of atoms, to one part in a million, and provided we use the correct renormalization procedure, no nasty infinite quantities appear. It clearly leaves open some questions, such as why are all values of electric charge integral multiples of the electron's charge (a phenomenon known formally as charge quantization), and why should the strength of the electromagnetic force, expressed by the fine structure constant, $\alpha = 1/137$, be the precise value it is? QED is clearly not a complete theory, but it is a very good theory and for a long time it has been the best theory that physicists have had at their disposal.

What of the weak force? By 1958 physicists had arrived at the V–A theory to describe the weak interaction. How does this shape up to our ideas of a 'good' theory? The V–A scheme, and Fermi's concept of a four-fermion point interaction on which it rests, is a phenomenological theory. It describes many of the phenomena of the weak interaction, and it describes them very neatly in a formalism with few unknown parameters and messy equations. But it is not a theory in the sense that QED is. As the closing paragraphs of chapter 5 revealed, V–A gives nonsense answers at high energies, and it is not renormalizable. Even with the introduction of the heavy intermediate vector boson, the 'photon' of the weak force, we still find infinite quantities in the calculations for some processes. A proper theory of the weak force needs something else, some deeper insight to allow it to be based on general principles, rather than descriptions of our observations.

Symmetry holds the key

The key to developing a respectable theory of the weak force turns out to be held not just in QED, but in Maxwell's theory of electromagnetism developed nearly a century earlier. An important characteristic of the electromagnetic field has to do with its symmetry properties, which conspire to conserve electric charge and make the photon the kind of particle it is. Now there are some obvious spatial symmetries in the effects of the electromagnetic force. The electric field round a point source of charge is spherically symmetric; a bar magnet's field is symmetric about the axis of the magnet (Figure 30). Moreover, there is a symmetry between positive and negative charge, and between north and south pole, so that one can be interchanged with the other and the fields still look the same. Which charge is positive and which negative, and which pole is north and which south are merely matters for definition. But electromagnetism contains more subtle symmetries than these, which are not so immediately obvious.

In electrical circuits all voltages, positive and negative, are referred to a zero level from which the others are measured. The world about us consists of matter that is electrically neutral, so we can define our zero level as 'earth' or 'ground' – literally, the voltage of the earth beneath us. But imagine a planet that is electrically charged relative to ours at some positive voltage. The inhabitants would perhaps know no better and still define the electrical zero level as that of their ground, and would measure all voltages as positive or negative with respect to this voltage, which *we* would find positive compared with *our* ground. What if the people on the planet performed similar experiments to those of Faraday, and Coulomb? They would still, in fact, find the same theory of electromagnetism as we do, for in Maxwell's equations it is the *difference* in voltage, or more strictly the difference in potential, that matters, not the absolute value. Maxwell's

Figure 30 *The electromagnetic force has some obvious symmetries: the electric field round a point charge is spherically symmetric (a), and the magnetic field of a dipole is symmetric about the magnet's axis (b)*

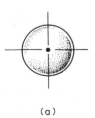

(a)   (b)

equations are in fact symmetric in the sense that you can shift the effective zero and still derive the same answers.

The kind of symmetry shown by Maxwell's equations is called a 'gauge symmetry', and we say that the theory is 'gauge invariant' because it remains unaffected by the 'gauge' changes. The word 'gauge' in this context arises from the work of Hermann Weyl in the 1920s. He considered the effects on electromagnetic theory of changes in the dimensions of space and time at different points in space–time – effectively changing the measuring rod, or gauge, of space–time. Weyl's attempts, which were intended to link electromagnetism with general relativity, were not successful, but the word he coined remains to describe changes made to a system at different points in space–time.

The symmetry we have imagined in the example of the planet at a net positive voltage is a particular kind of gauge symmetry – a global symmetry, where 'global' refers to the fact that we have imagined the same change everywhere, or globally. Maxwell's electromagnetism harbours a much more subtle symmetry though, a 'local' symmetry. This means that we can make different changes at different points in space–time, yet still end up with Maxwell's equations and the same theory of electromagnetism. Such symmetry on a local scale has to do with the interweaving of electric and magnetic fields, a change in one producing a change in the other. A theory with electric fields alone would not exhibit this local symmetry; it would not be invariant under local gauge transformations.

What happens to the gauge invariance, to the symmetry of electro-magnetism at the quantum level? Quantum electrodynamics also shows local symmetry, but with certain provisos. Maxwell's equations are concerned with electric and magnetic fields; QED is in addition concerned with the 'matter' fields describing charged particles, such as electrons, as we saw in chapter 4. It turns out that a quantum theory of the electromagnetic interactions of matter fields is locally gauge invariant only if the theory includes a field of infinite range with a quantum of spin 1 – a vector field. Such a theory is nothing more than QED, which incorporates the electro-magnetic field, with its massless photon which gives the field infinite range, and which has a spin of 1. Thus local gauge invariance forces us to introduce a field to mediate the interactions of charged particles, and this field brings with it the massless photon. A theory of particle fields alone cannot, it turns out, be gauge invariant.

So QED, which is such a good, well-behaved theory, exhibits local gauge invariance; indeed, it is this very property that makes the theory renormali-zable – free from impossible infinities. It does not seem too unreasonable, therefore, to conclude that local gauge invariance holds some kind of pointer as to what makes a good theory and what does not. The case for gauge invariance grows stronger when we realize that Einstein's theory of general relativity shows the same kind of local symmetry, which is imposed on space–time by the gravitational field. Thus two of nature's forces seem to exhibit the same symmetry; what of the others, the two that control the nature of the atomic nucleus?

Chapter 3 points out how the proton and the neutron are remarkably alike. Were it not for the electromagnetic force, they would be identical; the strong force makes no distinction between the two. This similarity, and that between other groups of particles with different charges, such as the pions, π^+, π^-, π^0, is expressed in terms of the property called isospin. The nucleon (proton or neutron) has isospin $I = \frac{1}{2}$, and the so-called 'third component', I_3, identifies two possible states: the proton has $I_3 = +\frac{1}{2}$, and the neutron has $I_3 = -\frac{1}{2}$.

In 1954, Chen Ning Yang and Robert Mills, at the Brookhaven National Laboratory, published a paper on the idea of a local gauge theory of the strong force. They proposed that isospin symmetry is the symmetry of gauge invariance for the strong force; that we can make local decisions about which particle is a proton and which a neutron, say, without altering the basic physics. To preserve the gauge invariance, as in the case of the quantum theory of electromagnetism, Yang and Mills needed to introduce a field – in fact, more than one field, as the symmetry of isospin is more complicated than the symmetry of electromagnetism.

The Yang–Mills theory, which was also developed independently by Robert Shaw at Cambridge University, has proved to be vitally important in our present understanding of the nuclear forces, although it was not particularly successful in its original aims. The theory does not provide a realistic formalism for the strong interactions because in its purest form it simply does not mesh with real life. In particular, the theory requires six vector (spin-1) fields of infinite range, analogous to the electromagnetic field, to preserve isospin symmetry locally, when different changes are made from one place to the next. Two of these fields can be associated with the electric and magnetic fields of the photon, but that leaves four more fields, describing two further massless quanta of spin 1.

The problem with the original Yang–Mills theory is that the two extra quanta are electrically charged, and there is no evidence whatsoever that charged photons can exist. Moreover, as Yukawa had argued more than twenty years previously the quanta of the nuclear forces have to be heavy to restrict the forces to the confines of the nucleus. It seems that, as far as the strong force is concerned, nature behaves with only approximate symmetry of the kind expressed in the Yang–Mills theory. The symmetry is spoilt by the necessity for heavy quanta.

Yang and Mills considered the strong force and isospin, but their general techniques were open to apply to other forces with similar symmetry. We can, for instance, picture the weak force as changing a particle from one equally likely state to another, as when a neutron decays to a proton (Figure 23). At the same time, the neutrino converts to an electron, and we can imagine this part of the process as a transformation between two partners in another doublet of particles, analogous to the neutron–proton doublet. Moreover, we can similarly couple the muon with its neutrino. So, although the 'hadronic isospin' associated with the strong force is not a good quantum number to consider in weak interactions, the symmetry of isospin does appear to have some relevance to the weak force. We can in fact identify a new quantum number, 'weak isospin', which characterizes all particles interacting via the weak force, the leptons as well as the hadrons.

Accordingly, in 1957 Julian Schwinger, at Harvard University, began to work on the weak force, basing his ideas on the theory of Yang and Mills. However, inspired by the analogies between the weak force and electromagnetism that arise in Fermi's theory, he included the electromagnetic force. The result is a theory that again has three massless vector quanta, one uncharged and two charged. As Yang and Mills had done, Schwinger identified the uncharged quantum with the proton, and regarded the charged quanta as the carriers of the weak force – after all, the weak force changes charge, as when the neutral neutron decays to a positive proton.

To bring the theory into line with reality, Schwinger had to endow these charged quanta with mass, which he did by allowing them to interact with additional fields. Moreover, because the theory related the weak quanta to the photon, it implied that the strengths of the weak and electromagnetic forces are in principle the same, echoing some work thirty years previously by Oscar Klein. Thus arose an interesting new idea: that the weak and electromagnetic forces are in a sense symmetric – different aspects of one and the same thing – but the weak force is weaker because its field quantum is heavy.

Schwinger's theory did not have the right prescription for the weak force; only the following year Richard Feynman and Murray Gell-Mann were to show that the weak interaction behaves in the 'vector-minus-axial-vector' manner of their V–A theory. Sidney Bludman, at the Lawrence Radiation Laboratory in Berkeley, California, did invent a gauge theory incorporating the correct V–A interaction, in 1958. His theory, unlike Schwinger's, was for the weak interaction only, but again being based on the formalism of Yang and Mills, it demanded the existence of three massless field quanta. In this case Bludman did not identify the neutral particle as the photon, but suggested that it was responsible for another kind of weak interaction, in which charge did not change. This 'neutral current' was carried, like the charged currents, by a heavy quantum, and as Schwinger had done,

Bludman had to modify his theory to give masses to the neutral and charged field quanta.

There were problems even with a theory like Bludman's that made it unsatisfactory. The theory could not predict what the masses of the carriers of the weak force should be. Instead, the masses had to be imposed on the theory in a rather arbitrary way to make it conform to nature. And there was the old problem of renormalization. Whereas the troublesome infinities of QED could be disposed of rather neatly, the infinities in the weak interaction simply would not go away. The introduction of masses always spoilt the gauge invariance and thereby removed the possibility of renormalization.

One of Schwinger's students, Sheldon Glashow, was concerned about the infinities associated with the weak force. He suspected, however, that the gauge theories held the key to eliminating the infinities, and that the right theory would have to involve the electromagnetic interaction as well. 'We should care to suggest that a fully acceptable theory of these interactions may only be achieved if they are treated together,' comments Glashow in his doctoral thesis of 1958. Three years later, Glashow published the first of a series of papers by a number of authors that began at last on the right track towards a 'unified gauge theory' of the weak and electromagnetic forces.

Schwinger's theory had been based, like the Yang–Mills theory, on the symmetry group SU(2) – the special unitary group in two dimensions. This is the technical name for the mathematical group that contains the operations, or transformations, under which the system is symmetric. The group must contain all the appropriate transformations for changing the system from one symmetric state to the next, as when you rotate a snowflake through 60° (Figure 31). It turns out that SU(2) is the appropriate group to describe weak isospin symmetry; the transformations that the group contains correspond to changes from one weak isospin state (a proton, say) to another (a neutron) by operations analogous to rotating a snowflake. (We can in fact talk of rotations in 'isospin space' between the neutron and proton states.)

Glashow's crucial step was to realize that the correct group to describe weak *and* electromagnetic interactions contains more transformations

Figure 31 *Snowflakes exhibit a six-fold symmetry of rotation. Turn a snowflake through an angle of 60° and it will appear the same as before; repeat this procedure six times and the snowflake will be returned to its original position. A symmetry group contains all the transformations – such as rotations – that change an object from one symmetric state to another*

than does SU(2). By analogy with some work on the strong force, he settled for the group called SU(2)xU(1), which is a combination of SU(2) and the unitary group U(1).

This extension of the initial SU(2) symmetry of Yang and Mills leads to a gauge theory with *four* spin-1 particles, the quanta of vector fields. Two of these are charged, and correspond to the weak charge-changing interactions such as occur in beta decay. The other two particles are neutral, one corresponding to the photon and the other, which Glashow called B, to a neutral, non-charge-changing weak interaction, similar to that proposed in Bludman's theory. At the time there were no compelling reasons to accept Glashow's suggestions. There was no evidence from experiments that a 'neutral-current' weak interaction existed and, to make matters worse, against all Glashow's hopes, the theory was not renormalizable.

However, across the Atlantic at Imperial College, London, Abdus Salam and John Ward had in fact come upon what was basically the same theory. Unlike Glashow, who had been trying to find a renormalizable, infinity-free theory, they had been trying to discover a theory for the weak interaction that was gauge invariant, a theory that was symmetric with respect to local variations in space–time. Here lies, with hindsight, a clue that Salam, Ward and Glashow were indeed on the right track. From rather different starting points and with different aims, and unaware of each other's work, they had all arrived at the same answer.

There were still some big hurdles to overcome. The theory as it stood did not give the masses of the field quanta, which still had to be introduced in an *ad hoc* fashion. Nor was it renormalizable. The plague of infinities afflicted this theory, as it had done all other attempts, because the two problems are related: it is in giving masses to the quanta to mimic reality that the infinities appear. The basic, unadulterated SU(2)xU(1) theory produces a symmetry between electromagnetic and weak interactions which is reflected in the zero masses of the four vector field quanta. Forcing masses upon some of these particles destroys this symmetry; it destroys the gauge invariance. The theory could, of course, simply be wrong; but there is an alternative to throwing it away completely. Suppose there is a mechanism, some way in which the quanta can acquire masses without destroying the gauge invariance, a salve that might cure the infinities as well.

Breaking the symmetry

Fortunately for the theory, there is such a mechanism, although it took several years for those physicists working on gauge theories to appreciate its tremendous value, for it first emerged in the rather different context of superconductivity – the state in which a material has zero electrical resistance at very low temperatures. The main relevant point to come out of the work on superconductivity is that the lowest energy state of a system – the ground state – can be asymmetric even if the system shows a symmetry at higher energies. The symmetry is said to be 'spontaneously broken' at the lowest energy, but this gives a misleading impression. A better description is to say that the symmetry becomes 'hidden'.

We know, for example, that we live on a roughly spherical planet, which creates a symmetric gravitational field; gravity pulls apples inwards towards the centre of the planet whether they fall off trees in Cambridge-shire in England, or in orchards in New Zealand (Figure 32). But ask a person in either country to say which direction is 'up' and which 'down' and you will receive conflicting answers relative to the background of 'fixed' stars. 'Up' in England is roughly in the direction of Polaris, the Pole Star; in New Zealand it is in the direction of the rather faint star known as σ Octantis, in the constellation Octans. From the point of view of each person, the symmetry is hidden; 'up' and 'down' define specific directions, whereas in reality they are only relative.

The idea of 'symmetry breaking' concerns changing from a state in which the symmetry of a system is obvious or overt, to a state where it becomes hidden. An astronaut finds that there is no preferred direction in space – once released, his apple will move off any which way, down to his feet, up to his head, or anywhere between. Down on earth, however, the symmetry is hidden from him, and 'up' and 'down' take on precise meanings.

Now the idea of hidden symmetry has been around since Werner Heisenberg pointed out that a piece of iron once magnetized exhibits

Figure 32 *The earth is almost spherical and its gravitational field approximately spherically symmetric. Thus gravity pulls apples 'down' towards the centre of the earth. But the absolute direction 'down' is different at different points on the globe. To a person in an orchard, gravity provides a unique sense of direction – 'down' is towards the ground – and the symmetry of the gravitational field is 'hidden'*

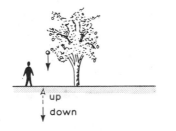

anything but symmetry, always preferring to point north–south, although the equations describing its magnetism are quite symmetric. What proved important to the gauge theories of the weak and electromagnetic forces was that hidden symmetry in a superconductor is related to the appearance of mass within the theory describing the superconductor's behaviour. The apparent asymmetry of the ground state of lowest energy is directly related in a rather subtle way to the generation of this mass; at higher energies where the symmetry is unobscured, the mass disappears.

A number of theorists working on field theories of elementary particles in the mid-1960s took over the ideas of hidden symmetry and began to incorporate them into their own work. In particular, Peter Higgs, of Edinburgh University, and F. Englert and Robert Brout, from the University of Brussels, found that they could use the technique to introduce masses into the theory of Yang and Mills without spoiling the gauge symmetry. The key lies in the fact that the symmetry is not destroyed, but merely hidden in the lowest energy state, when the masses appear. In other words, the theory with massive vector bosons is still gauge symmetric, although, with the appearance of the masses, this symmetry is no longer obvious.

The theory with masses describes the lowest energy state, the ground state of the system being considered. The so-called Higgs mechanism, which introduces the masses, does so by introducing a new field, with the unusual property of having least energy not when the value of the field is zero, but when it is above zero. Most fields, like the electromagnetic field, have least energy when the field is zero, but not so the Higgs field. This means that in the ground state – the lowest energy state – the Higgs field provides, in a way, a sense of direction in space, rather as the earth's gravity does for the astronaut once he comes down to land. This sense of direction 'breaks', or hides, the underlying symmetry of the theory in the ground state.

Only in 1967 did Steven Weinberg at Harvard University, and in the following year Salam, who was working independently, put the idea of hidden symmetry and the Higgs mechanism together with the $SU(2) \times U(1)$ theory of weak and electromagnetic interactions. Once they did this, the theory took on new life and produced values for the masses of the three vector particles of the weak interaction – the two charged particles, that is, W^+ and W^-, and the neutral particle which Weinberg dubbed Z^0. The values of the masses appear in terms of the so-called Weinberg angle, θ_W, which gives a measure of quantum mechanical 'mixing' between the two neutral particles, the photon and the Z^0. All this was for the price of one extra particle, a 'scalar' (spin-0) particle brought in by the Higgs mechanism. The theory actually acquires four scalar particles corresponding to four Higgs fields, but three of these 'disappear' in giving mass to the W and Z^0 particles; because the photon has no mass, the fourth Higgs particle remains observable, with a mass unpredicted by the theory.

With the Higgs mechanism, it turned out that Salam and Weinberg had eased away the trouble with the masses of the intermediate vector bosons. Their theory now exhibited an intellectually appealing symmetry, that of local gauge invariance. Both Salam and Weinberg also believed that they had surely cured the related problem of the infinities; the theory should now be renormalizable. However, renormalization is not easy to spot; in a sense it involves looking at a problem in a particular way, in which all the infinities cancel, rather like finding the right viewpoint to superimpose two objects that are at separate locations. Salam and Weinberg both stopped short of proving their theory renormalizable and turned to other things, but in Utrecht, Martinus Veltman and his student Gerard 't Hooft, were systematically studying the problems in renormalizing the Yang–Mills theory in which some infinities seemed stubbornly to remain.

In 1971, 't Hooft, inspired by some work on hidden symmetry, decided to investigate what would happen in the Yang–Mills theory if he included the Higgs mechanism to hide the local symmetry in the way reality seemed to require. 't Hooft had in effect hit renormalization right on the head: as he studied more and more complex reactions, all the infinities continued to cancel exactly. He had turned a good idea into a good theory, and as the theory of Weinberg and Salam was based on the Yang–Mills prescription, he had given this theory too its missing cloak of respectability. The Weinberg–Salam theory of the electromagnetic *and* weak forces at last emerged head and shoulders above all other putative theories of the weak force, although the time was still not ripe for it to displace V–A theory in the textbooks.

Introducing quarks

There was one important omission in the original work of Weinberg and Salam, which was soon rectified. Both had considered only the interactions of leptons, particles such as the electron and neutrino, that do not feel the strong force. But a complete theory of weak and electromagnetic interactions must clearly incorporate the strongly interacting hadrons, just as Fermi's theory and later V–A had done, so as to deal with processes such as neutron beta decay. In the mid-1930s, the proton and neutron were regarded as elementary particles, and Fermi had treated them as such in his theory of beta decay. By the early 1970s, however, there was good experimental evidence in favour of the ideas of Murray Gell-Mann and George Zweig, who proposed three fundamental 'quarks'

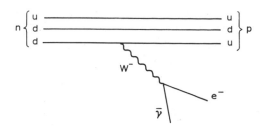

Figure 33 *The beta decay of a neutron to form a proton is seen at a more fundamental level as the transmutation of a constituent d quark, to a u quark, through the emission of a W⁻ particle*

as the basic constituents of all the known hadrons, as described in chapter 3.

With the quark model, an interaction between protons and neutrons becomes more complex than the simple analogy of colliding subatomic billiard balls. It is bunches of quarks that collide, and it is the reaction of individual quarks that moulds the observed form of the interaction. Fermi had regarded beta decay as the transmutation of a neutron into a proton, with the emission of a positron and neutrino; but this is clearly not the whole picture. In terms of quarks, we can visualize beta decay as the transmutation of one of the three quarks in a neutron, in such a way that it transforms the particle into a proton (Figure 33).

In 1960 and 1961, before the concept of quarks had emerged to inspire the whole community of particle physicists, Glashow had come across some problems in extending his theory of weak and electromagnetic interactions to include hadrons. Specifically, the problems arose with weak interactions that change strangeness – the property that was introduced to explain the production and decay of many of the new particles observed in the 1960s.

According to Glashow's theory, it should be possible for a reaction to occur in which strangeness changes, but electric charge does not. For example, a positive kaon, with strangeness $S = +1$, should be able to decay into a positive pion, a neutrino and an antineutrino:

$$K^+ \rightarrow \pi^+ + \nu + \bar{\nu}.$$

In this reaction the hadronic current, carried from the kaon to the pion, has no effect on the electric charge but changes its strangeness by one unit: we can call it a 'strangeness-changing neutral current'. Alternatively, the K⁺ can decay to a neutral pion, a positive muon and a neutrino:

$$K^+ \rightarrow \pi^0 + \mu^+ + \nu.$$

In this case the hadronic current changes the charge of the hadron by one unit in going from the K+ to the π^0; in fact, the change in charge (ΔQ) is the same as the change in strangeness. We can call the current in this second reaction a 'strangeness-changing charged current'. Now, experiments in the 1950s and early 1960s showed that although the first reaction does occur as frequently as might be expected, the second reaction occurs at only one hundred-thousandth of the rate, or only 1 in 10^5 times as often. Any viable theory of the weak force must therefore include some means of ensuring that strangeness-changing neutral currents do not arise.

In 1970, Glashow together with John Iliopoulos and Luciano Maiani, who were all at Harvard at the time, solved the difficulty, but at a price. They showed that they could cancel out the effects of strangeness-changing neutral currents provided they introduced a fourth kind of quark, in addition to the three that Gell-Mann and Zweig had proposed in 1964, and which were all that seemed necessary at the time to account for all the known particles. Contributions to the calculations from interactions between quarks of the fourth type and the others cancelled out the unwanted effects.

The three theorists, echoing some earlier work by Glashow and James Bjorken, referred to the fourth quark as carrying a new property called 'charm'. This distinguished the fourth quark from the others, and would contribute to the character of any particle the new quark formed, in the way that the strange quark endows particles such as kaons with strangeness. Any theory of the weak force would, according to Glashow, Iliopoulos and Maiani, have to incorporate the charmed quark if it was to describe correctly the decays of strange particles. Their work, of course, applied to the theory of Salam and Weinberg as much as to any other, and it provided the route to incorporating the hadrons – the particles made of quarks – in the same theory.

Thus by 1971 a theory of the weak interactions had emerged that was almost fully deserving of the title. It was based on general principles: on the ideas of symmetry in nature, of gauge invariance, of mathematical groups. Moreover, it was sensible in that it was free from unrealistic infinities. It involved relatively few 'free parameters' – quantities to be fixed purely by measurement and comparison with experiment. And, like any good theory, it made predictions that could be tested. Admittedly, it rested on the existence of a number of unobserved particles, the Ws and Z^0 and the Higgs, and the existence of a fourth quark, carrying a new quantum number, charm. But the picture was wonderfully self-consistent; it seemed it had to be true. It was now up to the experimenters to prove it to be true – or alternatively to disprove it.

7

Steps in the Right Direction

In the first few weeks of 1973, a single example of a very feeble reaction created an inordinate ripple of excitement among particle physicists throughout the world. A photograph of tracks in a bubble chamber revealed the path of a single electron, together with a few additional curls from other electrons knocked out of atoms in the chamber's liquid. That one apparently unremarkable image should create such excitement might seem surprising, but what the image held was a crucial piece of evidence, never previously observed, in favour of the 'electroweak' theory – the theory of weak and electromagnetic interactions that had only recently provoked interest through a newfound level of respectability. The image revealed the 'elastic' interaction of a neutrino with an electron, in other words a collision from which the neutrino and the electron emerge unscathed, having simply shared energy and momentum according to the appropriate conservation laws, as in a collision of billiard balls. A neutrino, being electrically neutral, leaves no observable track in a bubble chamber for it cannot ionize the liquid through electromagnetic interactions in the way an electron can. So the image showed but a solitary track – that of an electron, knocked out from its site within an atom in the liquid and propelled forwards by the impact of a high-energy neutrino.

The crucial image was one of many thousands taken of neutrino interactions in Gargamelle, a huge French-built bubble chamber located at CERN, Europe's international centre for particle physics near Geneva. Gargamelle was aptly named after the mother of Gargantua, the greedy royal giant who figures in François Rabelais's sixteenth-century satire, *Gargantua and Pantagruel*. The chamber, which went out of operation when it cracked in 1978, held 12 cubic metres of liquid freon (CF_3Br), a particularly suitable target owing to its high density of protons and neutrons. It was designed essentially as a long cylindrical box, 1.9 m in diameter, and installed with its axis aligned with the direction of the incoming neutrinos. This arrangement presented nearly 5 m of liquid to the neutrino beam, which meant that any particles produced could be tracked for a fair distance, thus providing opportunity for better identification.

The neutrinos entering Gargamelle came from the decays of high-energy pions and kaons, themselves produced in the interactions of protons with a metallic target. In the early 1970s the protons were provided by CERN's 28–GeV* proton synchrotron, which was at the time the laboratory's largest proton accelerator. Once accelerated, the protons were directed onto the target, producing many kaons and pions, among other particles. A special focusing system guided the pions and kaons in the same direction, and as they travelled along they decayed into muons and neutrinos, according to the reactions:

$$\pi^+ \rightarrow \mu^+ + \nu_\mu; \pi^- \rightarrow \mu^- + \bar{\nu}_\mu$$
$$K^+ \rightarrow \mu^+ + \nu_\mu; K^- \rightarrow \mu^- + \bar{\nu}_\mu.$$

Note that the neutrinos are specifically those of muon type, and that both neutrinos (ν_μ) and antineutrinos ($\bar{\nu}_\mu$) can be produced in this way, depending on the charge of the initial particles. To create a beam of neutrinos only, for example, you select the positive particles at the focusing stage.

The beam, now containing neutrinos and muons, as well as other particles, passed on for 22 m through many thousand tonnes of steel, so as to absorb all but the neutrinos; they interact so feebly with matter that they barely notice the earth, and even a large amount of steel has very little effect. Once through the steel, the neutrinos entered Gargamelle, lying in the embrace of a magnetic field which guided charged particles on curved tracks through the chamber.

The very weakness of the weak force makes reactions between neutrinos and other particles very rare, but Gargamelle's huge size at least increased the likelihood that some of these rare events might occur in the chamber's interior. Calculations suggested that in order to observe the elastic collisions of a neutrino (or an antineutrino) with an electron, as many as one million pictures would be needed, where every picture in principle 'observed' the arrival of some thousand million neutrinos derived from one pulse of protons from the main accelerator. By 1973 the team working on Gargamelle – comprising over fifty physicists from Aachen, Brussels, CERN, Paris, Milan, Orsay and London, who were backed up by scores of diligent film-scanners – had systematically worked their way through over half a million pictures: 375,000 from neutrino bursts, 360,000 from antineutrinos.

They struck lucky, for one picture showed the desired reaction, with an electron emerging as if from nowhere, travelling forward in the same

* See Appendix II for an explanation of GeV.

general direction as that of the incident antineutrino beam. Out of the 10^{15} antineutrinos that had travelled into Gargamelle, one had apparently produced the all-important reaction:

$$\bar{\nu}_\mu + e^- \rightarrow \bar{\nu}_\mu + e^-.$$

Why was this single event so important? The reaction is a weak interaction and, moreover, it involves no change of charge between the particles involved. It must therefore occur via the influence of a weak neutral current, in other words, through the exchange of the Z^0 particle predicted in the electroweak theory of Glashow, Salam and Weinberg (Figure 34a). Before 1973, there had been no evidence that such a reaction would occur. The discovery of the single event in Gargamelle suggested that it might after all occur, for the physicists were careful to rule out any other possibility. But, of course, more evidence was needed.

Elastic neutrino–electron scattering is so rare that even after scrutinizing 1.4 million pictures the team had discovered only three likely events. For better evidence of neutral currents they needed another reaction that could take place in Gargamelle more frequently. There is indeed such a reaction, but it is more difficult to interpret in terms of the electroweak theory as it involves the complex structure of nucleons – protons and neutrons. A muon–neutrino can interact with a nucleon, N, to produce a muon and a number of nuclear fragments, typically a nucleon and some pions (Figure 34b). This process can be written as:

$$\nu_\mu + N \rightarrow \mu^- + X,$$

where X represents the nuclear fragments, or hadrons – strongly interacting particles composed of quarks. Such a reaction is charge-changing, for the lepton which at first is neutral (the neutrino) becomes charged (the muon). However, an analogous reaction should be possible via the neutral current, in which case a neutrino rather than a muon should emerge (Figure 34c):

$$\nu_\mu + N \rightarrow \nu_\mu + X.$$

The physicists working on Gargamelle looked for both kinds of reaction, for the ratio of the number of events of each type is important in the electroweak theory. In both cases the appropriate image would reveal a shower of short tracks corresponding to the strongly interacting particles; but whereas the charge-changing process would produce the long virtually straight track of a high-energy muon, the neutral current would show nothing more, for the outgoing neutrino would leave as inconspicuously as the incident ones arrived. The length of Gargamelle was a vital factor in allowing the team to identify the particles correctly and pick out the appropriate events. After analysing 83,000 neutrino pictures and 207,000 anti-

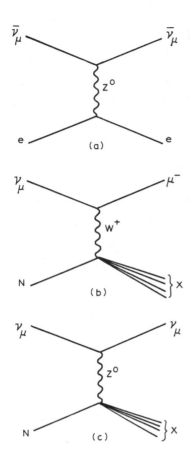

Figure 34 *The elastic scattering of a muon antineutrino ($\bar{\nu}_\mu$) and an electron (e) is a weak interaction involving no change of charge (a). According to electroweak theory, it occurs through the exchange of a Z^0 – the weak neutral current. The interaction of a muon neutrino with a nucleon (N) can take one of two courses: it can occur via the exchange of a W^+ – a charged-current interaction (b) – in which case a muon appears along with the nuclear fragments (X); or it can proceed via the weak neutral current (c), with the exchange of a Z^0, but in this case an undetectable neutrino emerges*

neutrinos, they found 102 examples of neutral currents from neutrinos and sixty-four examples from antineutrinos; the numbers of charged-current events were 428 and 148 respectively.

These results could be compared with the ratios of neutral-current to charged-current reactions predicted by the theory of Glashow, Weinberg and Salam. The theory sets the intrinsic strength of the weak force equal to that of the electromagnetic force, as embodied in the size of e, the charge of the electron, and the basic unit of charge. The weak force is much weaker only because its carriers, the W and Z^0 particles, are heavy, while the photon is massless. The main 'free' parameter which the theory does not fix is the so-called Weinberg angle, θ_W, which relates the amount of charged weak current to neutral weak current. It is the square of the sine of this angle ($\sin^2 \theta_W$) that can be derived from experiments. Those first results from CERN in 1973 gave a value for $\sin^2 \theta_W$ of between 0.3 and 0.4. Later experiments collecting much more data were able to improve on this figure. Moreover, the results from Gargamelle were soon confirmed by other experiments at the Brookhaven National Laboratory in New York, and at much higher energies at the new 300-GeV accelerator at the Fermi National Accelerator Laboratory (Fermilab) in Illinois. While at first it had seemed that the observations at CERN could be open to other interpretations, physicists soon accepted that the neutral currents of the electroweak theory were in all probability responsible for the events first revealed in Gargamelle's vast interior. The stage was set, and particle physicists were on the track to prove or disprove the electroweak theory.

The November revolution

If the discovery of neutral currents in 1973 made work with neutrinos highly fashionable, and was to guide the direction of such work for the next decade, then 1974 became destined to be the year to throw the spotlight onto experiments with colliding beams of high-energy electrons and positrons. While investigating some unusual behaviour in their studies of electron–positron collisions, a team of physicists from the Stanford Linear Accelerator Center (SLAC) and the Lawrence Berkeley Laboratory in California, hit upon something truly amazing. With the colliding beams of the SLAC's machine SPEAR turned to a narrow range in total energy, on Sunday, 10 November, the experimenters saw the rate at which hadrons were produced in the collisions rise to 100 times its usual value. Members of the team were quick to realize that what they were seeing was the creation and almost immediate decay of a new particle, with a mass some three times that of the proton, as revealed by the total energy of the colliding beams. The team, led by Burton Richter, christened the new particle the psi (ψ), and immediately began to draft a paper to *Physical Review Letters* to announce the discovery.

However, only the following day, Richter and his colleagues learned from Sam Ting, from the Brookhaven National Laboratory, that his team too had discovered a new particle, which had first shown up in August as an enhancement in the production of electron–positron pairs when high-energy protons collided with a beryllium target. In this case the total energy of the electron and positron revealed the mass of the short-lived particle. It was immediately clear that the two groups had stumbled upon the same thing, which Ting's group had named the J particle, after the Chinese character for 'Ting'. Within the next couple of days the editorial offices of *Physical Review Letters* received two papers, one from Ting's group and one from Richter's; two weeks later the results appeared in print, recorded for posterity.

One remarkable feature of this new particle, apart from its heaviness and the fact that it did not fit into the contemporary predictions of the quark model, which regarded all hadrons as built from three kinds of quark, was that it was relatively long-lived. This 'long' lifetime – only 10^{-20} s but still 1000 times longer than is typical of a heavy hadron, which should decay easily to lighter particles – is reflected in the narrow width of the peak corresponding to the particle on an energy plot (Figure 35). Such a plot shows the number of electron–positron collisions that produce hadrons, for example, versus the total electron–positron energy. The energy of the peak corresponds to the mass of the particle; the width corresponds to the lifetime, according to the dictates of Heisenberg's uncertainty principle (see chapter 4). Thus the shorter the lifetime of the particle, the more variation there can be in the energy used to create it, and the broader the peak. Conversely, the longer the lifetime, the less the amount of energy can vary, and the narrower the peak.

Figure 35   *Data from the two experiments that discovered the J/ψ particle. At both Brookhaven (left) and SLAC (right), the particle appeared as a resonance, or peak, in the production of e^+e^- pairs at a narrowly defined energy, equivalent to the particle's mass. (From, left: J. J. Aubert et al., Physical Review Letters, vol. 33 (1974), p. 1404; right, J. E. Augustin et al., ibid., p. 1406)*

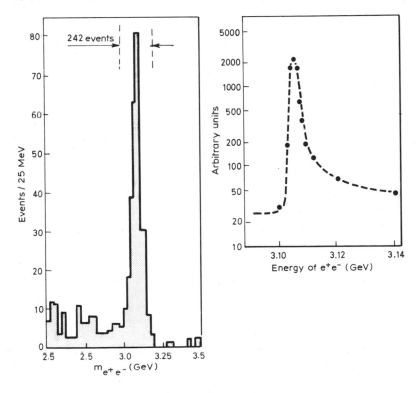

The width of the peak for the J/ψ, as well as other properties such as spin, provided the crucial clues to its nature. As described in chapter 3, it seems that the best explanation of the particle is that it is a meson, a bound state of a quark and its antiquark. But in this case, the particle had to contain a new kind of quark (and antiquark), for there already existed mesons corresponding to such states for the three quarks originally proposed by Murray Gell-Mann and George Zweig. The idea of a fourth type of quark, with a new quantum number, was exciting, but it was not without precedent. Recall from chapter 6 that Sheldon Glashow, together with Maiani and Iliopoulos, had required there to be a fourth quark in order to explain the absence of strangeness-changing neutral currents – weak interactions in which the total strangeness changes by one unit, but in which charge does not flit between leptons and hadrons.

Within the months following the announcements from Richter and Ting, the consensus grew that the most reasonable explanation for the new particle is that it is indeed the meson formed from a new 'charmed' quark, bound with its antiquark. Before the end of November, Richter's team had found a second particle, the ψ' ('psi prime'), slightly heavier than the first (Figure 36). While the J/ψ is formed from the charmed quark and antiquark (cc̄) in their lowest energy configuration, the ψ' corresponds to the first excited state. The two particles are in fact members of a whole family of cc̄ states, discovered in the fullness of time.

Figure 36 *Tracks in a spark chamber detector reveal the decay of a ψ' particle into a J/ψ and two pions. The J/ψ decays into an e⁺e⁻ pair, producing four tracks in all. (From G. S. Abrams et al., Physical Review Letters, vol. 34 (1975), p. 1181)*

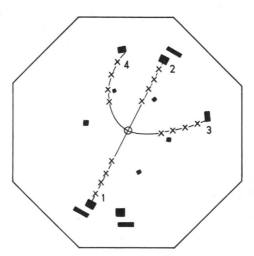

With the discovery of the J/ψ and ψ′ the hunt was on for other particles comprising a charmed quark (or antiquark) bound with one or two of the commoner quarks (or antiquarks) to form a meson (q$\bar{\text{q}}$) or baryon (qqq). Such particles would themselves carry the quantum or 'label' of charm, just as strange particles carry a quantum of strangeness, endowed upon them by the strange quark. In the J/ψ, on the other hand, the 'anticharm' of the antiquark cancels the charm of the quark, so that the net charm is zero. But particles with non-zero charm should exist if the hypothesis is indeed correct.

The search proved relatively slow, and it was not until April 1976 that the same team at SLAC found good evidence for the D^0 meson – a combination of a charmed quark (c) and an 'up' antiquark ($\bar{\text{u}}$). A couple of months later the team found signs of the related D$^+$ meson (c$\bar{\text{d}}$), and at last, with other possible sightings at CERN, Brookhaven and Fermilab, it seemed reasonable to believe that there are indeed (at least) four different types of quarks, just as Glashow, Iliopoulos and Maiani had demanded. Another piece was slotted into the electroweak jigsaw, and Ting and Richter received their Nobel prize in November of the same year.

The evidence accumulates

By 1976 the evidence for two of the missing ingredients of the electroweak theory looked good; weak neutral currents and charmed quarks were part of nature's grand design after all. The experiments that accumulated the evidence were by no means easy, and some were masterpieces of technique, which observed extremely subtle but none the less vital effects. One such experiment came to fruition at SLAC in 1978, this time using the famous 3-km-long linear accelerator, rather than the colliding-beam machine, SPEAR.

Charles Prescott and colleagues, mainly from Stanford and Yale Universities, were interested in studying the effects of the weak force in the interaction of electrons with quarks – in practice, the interaction of electrons with protons and neutrons. Now, electrons and quarks are electrically charged, so they interact via the electromagnetic force as well as the weak force. As we saw in chapter 5, the strength of the weak force, as characterized by the coupling constant, G_F, is some 1000 times weaker than that of the electromagnetic force, given by the fine structure constant, α. Moreover, the probability of an interaction depends on the *square* of the coupling constant, so weak interactions occur 1000^2, or one million, times less often than do electromagnetic effects. At first sight, then, the chance of being able to pick out effects due to the weak force in a situation dominated by the electromagnetic force seems nigh on hopeless (which, of course, is why so much work on weak neutral currents has been done with neutrinos which do not feel the electromagnetic force).

There is, however, one property of the weak force that distinguishes it from the electromagnetic force, and that is its disrespect for parity symmetry. As Lee and Yang predicted, and as Wu and her collaborators found, the weak force violates parity. In other words, it distinguishes between right- and left-handed interactions, so that the probability for a left-handed weak process is not the same as that for its right-handed counterpart. In terms of the electron–quark interactions that Prescott's team studied, this means that the weak interactions of left-polarized electrons (spinning like left-hand corkscrews about their directions of motion) are slightly different from those of right-polarized electrons. This difference is sufficient to show up in a comparison of the overall reaction rate – both electromagnetic and weak – for left-polarized and right-polarized electron–quark collisions.

Prescott's particular study was of collisions between electrons and deuterons, the nuclei of deuterium, or 'heavy' hydrogen, which contain one proton and one neutron. The choice of deuterium is important, for it provides a target with essentially equal numbers of 'up' quarks (u) and 'down' quarks (d). Many complex effects relating to the precise distribution of quarks in the nuclei tend to cancel out in calculations on deuterium, and it is relatively easy to compare the data with theoretical predictions. Using both left- and right-polarized electrons, Prescott studied the reaction:

$$e^- + D \rightarrow e^- + X,$$

where D stands for the deuteron, and X represents any combination of hadrons produced in the collision. The reactions are actually between an electron and one of the quarks in a deuteron, and can occur either via the electromagnetic force or the weak force. In the former case a photon is exchanged, in the latter a Z^0 particle (Figure 37). Note that it is the weak *neutral* current, predicted by the electroweak theory, that comes into play, for charge does not pass from the electron to the hadrons.

If $\sigma_R(\sigma_L)$ represents the probability for the reaction between right-(left-) polarized electrons and deuterons, then we can define an asymmetry, $A = (\sigma_R - \sigma_L)/(\sigma_R + \sigma_L)$, which is the difference in the probabilities, divided by their sum. From a knowledge of the strength of the weak and electromagnetic interactions, we can work out that A should be approximately $10^{-4} \times q^2$, where q is the momentum transferred from electron to quark in the interaction. This in fact meant that Prescott and

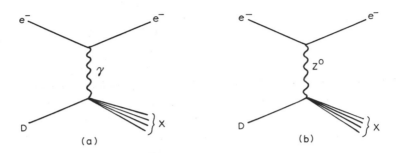

Figure 37 *When an electron (e) scatters inelastically off a deuteron (D) to produce a number of hadrons (X), the interaction occurs via one of two possible neutral currents: due to the exchange of a photon (γ) as in (a), or a Z^0 (b). The weak interaction involving the Z^0 violates parity conservation, and it creates an asymmetry in the scattering of polarized electrons from deuterons, which can be measured and compared with the predictions of electroweak theory*

his colleagues were looking for an effect occurring at the level of about 0.01 per cent or even less.

The electron beam accelerated in the linac at SLAC is normally unpolarized, so the team's first task was to produce an appropriate source of polarized electrons. Physicists from Stanford and Yale had already built a polarized source in 1971, but for the new experiment they developed a better source that would provide a beam of electrons as intense as the normal electron gun. The electrons were produced from a crystal of the semiconductor gallium arsenide, which emits electrons by the photoelectric effect. What is more, the electrons are either right- or left-polarized depending on the polarization of the incident light.

To 'pump' the electrons from the gallium arsenide, Prescott's team used pulses of circularly polarized light from a dye laser, emitting at a wavelength of 710 nm, in the red part of the spectrum (Figure 38). The laser actually emits linearly polarized light – vibrating in one direction perpendicular to its direction of motion. This was converted to circularly polarized light, in which the direction of vibration rotates either clockwise (right-polarized) or anticlockwise (left-polarized) as the light travels along. With right-polarized light, the gallium arsenide emitted electrons with an overall right-polarisation of 37 per cent; with left-polarized light the electrons were left-polarized to the same extent.

The polarization of the electrons could be varied at random from one pulse to the next, as the circular polarizer was set by a voltage, the sign of which (positive or negative) determined the polarization of the outgoing light. In addition, a linear polarizer – a calcite crystal – determined the linear polarization of the light entering the circular polarizer.

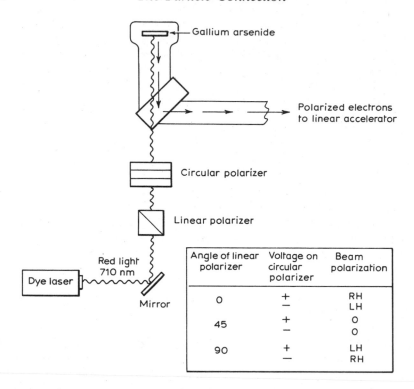

Figure 38 *The experiment at SLAC to measure the asymmetry in the scattering of polarized electrons by deuterium requires a special source to provide the polarized electrons. The combined settings of the two optical polarizers determine the ultimate polarization of the electron beam as the table shows*

With the crystal at 0°, the output of the circular polarizer was unaffected; at 90° right-polarized light from the circular polarizer was changed to left-polarized and vice versa; and with the crystal at 45°, the light was completely unpolarized, so the experimenters could test their system with unpolarized electrons.

One additional feature of this experiment was that, instead of measuring the reactions of individual electrons, the apparatus was set up to record the flux of electrons scattered during each pulse from the linac. This technique allowed the team to observe many more reactions than in a conventional experiment (which would track individual events) and it improved their chances of observing the minute effects of electroweak interference.

Using the linac to accelerate the polarized electrons to energies from 16 to 22 GeV, Prescott's team measured both the polarization of the electron beam and the number of electrons scattered from the deuterium target. And at a variety of energies they determined the asymmetry, A, with the calcite crystal at $0°$, $45°$ and $90°$. As expected, the results showed essentially no asymmetry with the crystal at $45°$, that is, with unpolarized light striking the gallium arsenide crystal, and therefore with unpolarized electrons (Figure 39). At other angles, though, an asymmetry did appear.

The effect of the weak neutral current – the exchange of the Z^0 – means that deuterons show a slight preference for scattering left-handed electrons over right-handed electrons, and the experiment showed that

$$A = -(9.5 \pm 1.6) \times 10^{-5} \times q^2,$$

where q^2 represents the momentum transferred from the electrons to the deuterons. By measuring A for different values of the fraction of energy the electrons lost to the hadrons, Prescott could also compare his data with predictions of the electroweak theory and obtain a measure of the parameter, $\sin^2 \theta_W$. The results show that $\sin^2 \theta_W = 0.22 \pm 0.02$, which is in good agreement with the latest data from neutrino experiments.

I have described the polarized electron source used by Prescott in some detail in order to show the elaborate lengths required to reveal the subtle

Figure 39 *The measured asymmetry in the scattering of left- and right-polarized electrons from deuterons. The prism (linear polarizer) effectively selects the polarization of the beam; at 45° the beam is unpolarized and no asymmetry is observed*

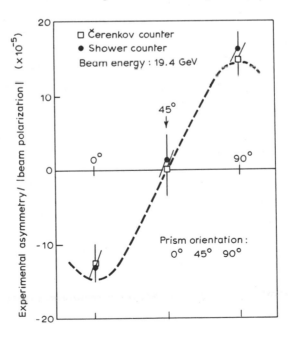

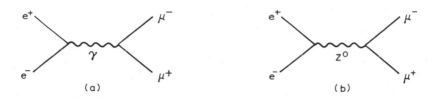

Figure 40 *The annihilation of an electron–positron pair (e⁺e⁻) can produce a muon pair (μ⁺μ⁻) either via the electromagnetic force and the exchange of a photon (a), or through the weak force and the exchange of a Z⁰ (b). The latter reaction produces a forward–backward asymmetry in the production of muon pairs, which can be measured and compared with the predictions of electroweak theory*

effects of the weak interaction, particularly when it is overshadowed by the much stronger electromagnetic force. In recent years studies of high-energy collisions between electrons and positrons have begun to explore further consequences of the 'interference' between the weak force and electromagnetism, which become apparent only at relatively high energies.

Experiments first at DESY, the German particle physics laboratory near Hamburg, and then at SLAC, have investigated the annihilation of electrons and positrons to produce muons:

$$e^+ + e^- \rightarrow \mu^+ + \mu^-.$$

Once again, this reaction can occur either via the electromagnetic force (photon exchange) or the neutral weak force (Z⁰ exchange), and again the electromagnetic interactions swamp the weak ones (Figure 40). However, interference between the two processes produces an asymmetry in the way the muon pairs (μ⁺μ⁻) are produced, which increases with energy and should be visible in the experiments conducted at DESY and SLAC. In this case, the asymmetry does not arise from parity violation, but concerns the direction in which the muons fly off from the reaction. Suppose we define the (major) part of the electron beam that does not annihilate as moving 'forwards' as it leaves the collision; then the positron beam carries on 'backwards' (Figure 41). It might seem that the μ⁻ that are produced can go either way, as long as the μ⁺ move the opposite way to conserve momentum and energy. However, according to quantum electrodynamics there should be a slight tendency, at the level of about 1 per cent, for the μ⁻ particles to prefer to leave in the 'forward' direction, that is, the same direction as the outgoing electron beam. But the neutral weak force should produce an even more dramatic effect, in which the μ⁻ prefer to travel 'backwards', in the direction of the positron beam. If

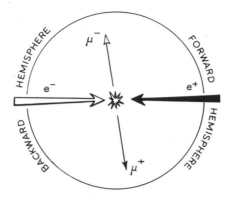

Figure 41 *In head-on collisions between beams of electrons (e⁻) and positrons (e⁺) we can define 'forwards' as the direction of the outgoing beam of electrons that have not collided. We can then measure the directions of muons (μ) produced in the collisions; if produced by the weak neutral current the μ⁻ should tend to emerge 'backwards'*

N_F and N_B represent the number of μ^- travelling into the forward 'hemisphere' and the backward 'hemisphere' respectively, we can define an asymmetry $A = (N_F - N_B)/(N_F + N_B)$, which is the ratio between the difference in the numbers of events in either direction, and their sum.

At PEP, the electron–positron collider at SLAC, with a total energy of 29 GeV, the combined results from three different detectors indicated an asymmetry of −6.4 ± 0.9 per cent, as of July 1983. This compares favourably with the value of 5.3 per cent predicted by the electroweak theory with sin² θ_W set at 0.23. PETRA, the collider at DESY, runs to higher energies than does PEP, and the results from four experiments on this machine can be combined to give a value for A of −13.2 ± 4.0 per cent, at a total energy of 40.3 GeV.* Again this figure compares favourably with a predicted value of −13.6 per cent. The energy is still not high enough at 40.3 GeV, and the asymmetry not large enough, to determine the mass of the Z⁰ exchanged in the weak interaction, but the data are certainly sufficient to show the behaviour expected from electromagnetic-weak interference. A combination of all the data at all energies provides a dramatic illustration of how the asymmetry increases with energy, and deviates from the prediction of quantum electrodynamics alone, in other words from the exchange of photons alone (Figure 42).

* As reported by Albrecht Böhm from DESY at the International Europhysics Conference on High Energy Physics in Brighton, July 1983.

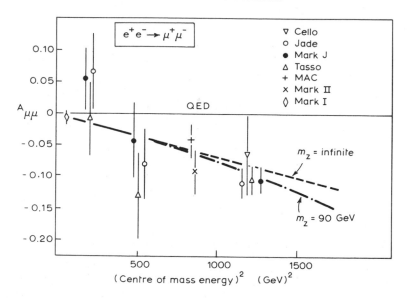

Figure 42 *Measurements of the forward-backward asymmetry in e^+e^- by a number of experiments show how the asymmetry increases with the total e^+e^- energy. The data diverge from the prediction of QED alone (no Z^0), suggesting that the Z^0 particle does exist, but the energy is not high enough to give a measure of the Z^0's mass. (From: A. Böhm, DESY preprint 82-084)*

Electron–positron annihilations into muons provide a good means of studying the weak neutral current without the complicating influences that arise in the production of quarks and their materialization as hadrons. As such, these studies rank with those of neutrino–electron scattering. One other area in which the influence of the weak neutral current should be manifest is in the realm of atomic physics. Here it is once again the parity-violating nature of the weak force that should reveal itself, over and above the effects of the electromagnetic force, acting between the atomic electrons and the quarks of the nucleus.

One kind of atomic physics experiment has aimed to detect the rotation of the polarization of laser light as it passes through a vapour of bismuth – an effect that is as small as a few hundred-thousandths of a degree. An alternative has been to study how the intensity of fluorescence from thallium varies as it is exposed to a beam of light of varying polarization. These kinds of experiment are exceedingly difficult and have in the past given different answers, some of which failed to reveal the influence of the weak neutral current. However, there now seems to be a growing consensus that the results are generally what might be expected from the electroweak theory.

The missing links

Taken together, the results from the variety of experiments that have explored the consequences of the weak neutral current provide fairly substantial evidence in favour of the electroweak theory. A comparison of all the data with theory results in a value for $\sin^2 \theta_W$ of 0.233 ± 0.009.[*] It is possible to relax some of the assumptions made in the theory, and so have more 'free parameters' to be determined by comparing predictions with results, but the answers come out to be remarkably similar to the values assumed in the theory, and lead to virtually the same value of $\sin^2 \theta_W$.

However, despite the discovery of neutral currents and the charmed quark, there is still something missing. What evidence is there that the Z^0 and its companion charged particles, the W^+ and W^-, really do exist, in the sense that they can be produced and observed to decay in experiments? Although the electron–positron asymmetry measurements could 'see' the influence of the weak force, they could not, at the energies available by the end of 1982, reveal the mass of the Z^0. This comes down to saying that the experiments could not observe the difference between Fermi's point interaction of four fermions (in effect the exchange of an infinitely massive Z^0) and the quantum exchange of the electroweak theory. Fortunately, the electroweak theory provides the very clue as to why no Z^0 or W particle had appeared in experiments up to 1982. One of the theory's strong points is that it predicts the values of the masses of the intermediate vector bosons, in terms of the Weinberg angle, θ_W. So from all the experiments on neutral currents, parity violation and so on, we can tell precisely what the mass of these particles should be if the theory is indeed correct. The theory says that: $M_W - 37.3/\sin \theta_W$ and $M_Z = 74.6/\sin 2\theta_W$, in units of GeV. (The proton's mass is a little less than 1 GeV.) Putting in the average measured value for $\sin^2 \theta_W$, and making some slight adjustments to allow for the complex processes, involving more than one Z^0 or W, that are encompassed in the renormalization of the electroweak theory, gives the following values: $M_W = (79.2 \pm 1.5)$ GeV; and $M_Z = (90.7 \pm 1.2)$ GeV.[†] Until 1982 there existed no particle accelerator capable of generating such massive particles – eighty and ninety times the mass of the photon respectively. From the first observation of the weak neutral current in 1973, it was the best part of a decade before the appropriate machine was ready to insert the next all-important pieces into the electroweak puzzle.

[*] J. E. Kim *et al.*, *Reviews of Modern Physics*, vol. 53 (1981), p. 211.
[†] G. Myatt, *Reports on Progress in Physics*, vol. 45 (1982), p. 1.

8

The Antiproton Solution

The electroweak theory of Glashow, Weinberg and Salam gained respectability in the eyes of theorists in 1971 when Gerard 't Hooft's work showed that it was possible to rid the theory of its embarrassing infinities. It gained respectability in the eyes of experimenters in the exciting years of 1973–74, when a variety of discoveries confirmed some of the theory's predictions and helped to strengthen convictions that electroweak unification was indeed a subtle symmetry of nature. Thoughts turned immediately to the W and Z^0 particles, the intermediate vector bosons that the theory required to carry the weak force between particles (or, strictly, particle fields) just as the photon carries the electromagnetic force. The electroweak theory predicted masses for these particles in terms of a parameter, $\sin^2 \theta_w$, that had to be derived from measurements, and it soon became apparent that the masses must lie in the range 50–100 GeV if the theory was correct. Since the mid-1970s, improved measurements have refined the predicted values for the masses, but even so it was clear that no existing particle accelerator could provide access to the appropriate energy regime.

In 1974, the largest accelerator was the proton synchrotron at Fermilab, near Chicago. With over 1000 magnets to steer the beam round a ring 2 km in diameter, this machine could accelerate protons to 400 GeV, which might on the face of it seem sufficient to produce particles of mass equivalent to a modest 100 GeV. However, there are some fundamental limitations with a machine like Fermilab's proton synchrotron and the similar device, the Super Proton Synchrotron (SPS), that came into operation at CERN in 1976. These machines accelerate protons to a maximum energy – 400 GeV, say – and then flick the particles out of the synchrotron's main ring of magnets towards experiments. The whole cycle takes several seconds and then new particles are injected.

The problem arises when a 400 GeV proton strikes the nucleus of an atom in a target, of metal, hydrogen, or whatever. The proton travels rapidly towards the target at velocities a few ten-thousandths of a per cent less than that of light. After the collision, the resulting debris is thrown onwards in the direction in which the incident proton was travelling. In other words, much of the proton's energy is transferred as kinetic energy to the products of the collision, rather as fragments of a target hit by a high-velocity bullet will continue in the bullet's direction. Only a small portion of the initial energy is left for the creation of new particles. Indeed, if E is the energy of the incident proton in GeV, then only approximately $\sqrt{(2E)}$ is not taken up as kinetic energy. So when a 400 GeV proton strikes a target, only about 28 GeV is in effect freed – a figure well below that of the mass suspected for the intermediate vector bosons.

We can see immediately from a graph showing how $\sqrt{(2E)}$ rises with E that increasing the maximum value of E is not an efficient route towards higher available energy – increasing E by a factor of 100 will increase the available energy by only 10 (Figure 43). Fortunately, there is another way to gain access to higher energies, and that is by shooting two high-energy beams of particles at each other to create head-on collisions. In this case virtually all the energy of the two colliding particles is free to take part in the interaction and can be used to create new particles. This is rather like the fact that a head-on collision between motor vehicles is far more devastating than when one vehicle drives into another that is stationary, although high-energy particles are travelling very close to the speed of light and the correct calculations must employ Einstein's special relativity.

In the mid-1970s there were already a few machines at high-energy physics laboratories that exploited the idea of head-on collisions. The one that reached the highest energy was at CERN, and collided two beams of protons together. The machine, known as the Intersecting Storage Rings (ISR), consisted of two interlaced rings, in which protons circulated in opposite directions.* The word 'storage' in the name refers to the fact that in such machines the particles are indeed stored, circulating many times,

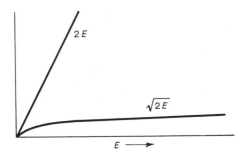

Figure 43 *When a beam of particles of energy E (in GeV) strikes a stationary target, only $\sqrt{2}$E is available for the creation of new particles and so on; and increasing E does not produce a large increase in $\sqrt{2}$E. With colliding beams, however, a total energy of 2E is available*

---

* The ISR was closed down in May 1984 to be dismantled.

because only a few particles actually collide when the two beams cross. For there to be a reasonable collision rate in experiments surrounding the cross-over points in the ISR, there needed to be far more protons circulating than there are in the beam of particles from the 28 GeV proton synchrotron (the PS) that fed the rings.

A beam of protons in the PS contains some 10^{13} protons concentrated into twenty bunches. The bunches are accelerated and then ejected towards their ultimate destination, for instance, the ISR. But the business of colliding particle beams is rather like firing shotguns at each other: collisions between the individual particles are rare. To reach a useful number of collisions, the ISR needed circulating beams of 4×10^{14} protons, over a factor of 10 more than in the beams supplied by the PS. Thus each ring of the ISR had to store together many bunches from the PS – a process known as 'stacking'. Once full with the appropriate number of protons, the ISR accelerated the beams to 31.5 GeV and kept them circulating at this energy for several days. The total energy of the proton–proton collisions in the ISR was thus 63 GeV – equivalent to the energy released when a beam of 1800 GeV strikes a stationary target.

But 63 GeV is not enough to create W and Z^0 particles if the electroweak theory of Glashow, Salam and Weinberg is correct – and indeed no signs of these particles were ever found in experiments at the ISR. Instead, particle physicists in the 1970s began to look to new machines which could be built with the potential to discover the missing particles. In particular, there was a plan to build a machine called Isabelle, a huge version of the ISR at the Brookhaven National Laboratory in New York. This would collide beams of protons at energies of 200 GeV per beam (later proposals increased this to 400 GeV), to give a total energy of 400 GeV (800 GeV), well above the energy for the direct production of intermediate vector bosons. Scheduled initially to start up in the early 1980s, the project, known as Isabelle, was beset with difficulties, particularly in the design of the magnets required to steer the beams ultimately to 400 GeV. To save on electrical power, the project was to use superconducting magnets – electromagnets with coils built from superconductor, a material with zero electrical resistance at very low temperatures. Eventually, in 1983, with the tunnel complete, but no magnets to put in it, the project was abandoned, mainly for reasons that will, no doubt, become apparent.

Meanwhile, in Europe, a number of ideas were under study, in particular to build a proton synchrotron to reach energies over 1000 GeV; to collide protons at 400 GeV per beam; and to collide electrons with protons at high energies. However, by 1976 plans developed to build a different kind of storage ring, one that collides high-energy electrons with their antimatter counterparts, positrons. Small machines of this type existed in the early 1970s, and began to prove their exceptional worth with the discovery in 1974 of the J/ψ particle in electron–positron collisions at SLAC. An electron–positron collider can be much simpler than a proton–proton machine – in principle only one ring of magnets is required. This is because the effect of a magnetic field on a positively charged particle is to bend it in the opposite sense to a negatively charged particle; this is the basis for the widely used technique for separating out particles of a specific charge. However, consider the paths of particles of opposite charge travelling round a ring of magnets: if the magnets bend negative particles to the right, so that they travel clockwise, the same magnets will bend positive particles to the left, so that they travel round the same ring of magnets in an anticlockwise direction (Figure 44). In this way electrons and positrons can be stored and collided in a single ring of magnets, which allows a great saving in electricity as well as materials when compared with the alternative of two interlaced rings of magnets. The two beams in fact travel on slightly displaced paths, and are made to intersect at predetermined locations where detectors surround the beam-crossing.

CERN's final choice for the future, which was not in fact fully approved until 1982, was to build a huge electron–positron collider, reaching ultimately 100 GeV per beam. The project has to be huge, because electrons and positrons radiate significant amounts of energy while travelling along curved paths – an effect that is much smaller for protons, because protons are 1800 times as heavy as electrons. By making the ring of magnets larger, the curves become shallower and less radiation occurs for the same energy. This in turn means that less electrical power is needed to compensate for radiation losses while accelerating the particles. The machine proposed for CERN, called LEP for Large Electron Positron machine, will be 27 km in circumference and will start up in 1987.

Figure 44   *Magnets that guide a beam of electrons in one direction around a synchrotron ring will guide positrons (with opposite charge to electrons) in the opposite direction. Thus a single ring can contain two counter-rotating beams of oppositely charged particles, one above the other, and these can be made to collide at predetermined points*

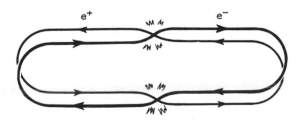

But patience does not come easily to scientists who feel themselves to be on the verge of a breakthrough. Back in the mid-1970s it was clear that large new projects such as LEP and Isabelle could not be built quickly, and that it would be the mid-1980s before the experiments could test out the electroweak ideas. There had to be a shortcut to the W and Z^0 particles, or at least to the energies at which they were expected to materialize. In 1976 Carlo Rubbia from CERN, together with David Cline from the University of Wisconsin and Peter McIntyre from Harvard University, proposed such a shortcut. Their suggestion was to convert the existing large proton synchrotrons into colliders, by feeding antiprotons into the magnet ring to circulate in the opposite direction to the protons. This solution would not only be quicker, but also cheaper, as it would employ many existing components.

How to catch antiprotons

The basic idea of a proton–antiproton collider is deceptively simple; in practice, it presents some tremendous challenges. First, you have to produce antiprotons, and produce enough of them for there to be sufficient proton–antiproton collisions to creat a reasonable number of W and Z^0 particles, if they exist in the way that theory predicts. The discovery of one W particle in an experiment that had run for a year would probably convince no one – the aim was to produce W particles at least at the rate of one a day.

Protons are relatively easy to produce, for they are simply the nuclei of hydrogen atoms. The basic principle is to excite the atoms in hydrogen gas with an electrical discharge so that the electrons break away from the nuclei – in other words, the gas becomes ionized. The positively charged protons can then be attracted towards a plate at negative voltage with a hole in it, through which a beam of relatively low-energy protons will appear. With antiprotons the task is more difficult, for the particles do not occur naturally. Instead they must be created in the collisions of high-energy protons with a target; the antiprotons can be separated out relatively easily from the other debris, using magnetic fields. In the collisions each antiproton must be produced in conjunction with a proton, so as to conserve the total number of baryons, for the antiproton is an *antibaryon*. So in a proton–proton collision the most simple reaction that can produce an antiproton is:

$$p + p \rightarrow p + p + p + \bar{p},$$

where the bar above the last p indicates the antiproton. The proton and antiproton have the same mass of nearly 1 GeV, so there has to be at least 2 GeV available in the collision for an antiproton to be produced together with the additional proton, and this in turn means that the incident proton must have an energy of nearly 6 GeV.

But the problems really only begin at the target, for the antiprotons are produced with a wide range of energies and directions, hardly ideal for injection into an accelerator that operates by keeping particles of precise energies on precise orbits. Introduce a beam of antiprotons directly from a target into an accelerator and most of them would soon be lost straying widely from the optimum orbit through the ring of magnets. Instead, the spread in momentum (the product, velocity × mass) of the initial beam must first be reduced; fortunately, there are a couple of different techniques available to do this.

The spread in momentum of antiprotons produced at a target can also be regarded as an effective temperature. From the point of view of an antiproton in the beam, the fact that the particles have a spread in momentum means that they appear like the molecules of a gas, jiggling around relative to each other. The more they jiggle, the higher the effective temperature, and the broader the spread in momentum relative to the average value. One technique to 'cool' antiprotons dates from 1966 when it was proposed by Gersh Budker, director of the Institute for Nuclear Physics at Novosibirsk in the Soviet Union, as part of a plan to collide protons with antiprotons at an energy of 25 GeV per beam. Budker's idea was to let the antiprotons mix for a while with another 'gas' at a lower 'temperature'. This gas would be a beam of electrons, all with more or less the same momentum, and therefore a low effective temperature. The two beams would have the same *average velocity*, but the electrons would have a much smaller *spread* in velocity, and hence momentum. Repeated encounters between the antiprotons and successive bunches of electrons would eventually cool down the antiprotons sufficiently. Budker's technique for 'electron cooling' was first shown to work in tests at Novosibirsk in 1976.

However, in 1975, tests on protons in the ISR had shown that another technique for beam cooling would work. The method used is called 'stochastic cooling' and was invented by Simon van der Meer at CERN in 1968 (although he did not publish his ideas until 1972). The aim is to apply many small corrections to the orbits of a random selection of the beam particles, so as to nudge them onto the optimum path and thereby gradually concentrate the beam. The name 'stochastic' reflects the random nature of the operation. Van der Meer suggested sensing the average position of a slice of the beam at one point in the accelerator ring. The signal from this sensor could then be used to apply an electric field to move the particles towards the correct path at a point farther

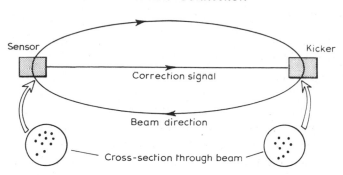

Figure 45   *In stochastic cooling, a sensing electrode detects the average distribution of particles in a slice through the beam. This signal is used to generate a 'kick' sufficient to push the particles* on average *to the correct position, although some will be pushed farther away. Repeated many times on slightly different slices, the net effect is to concentrate the beam around the correct orbit*

along the orbit (Figure 45). This 'kick' would be too much for some of the particles and not enough for others, but on average it would have the desired effect. On the next orbit of the machine, the sensor would operate on a slightly different slice of the beam, as the particles would all have moved relative to each other, and a somewhat different group of particles would be nudged towards the correct path. By repeating this procedure many times, the complete beam would gradually be concentrated, and the large initial spread could be brought under control.

The rigorous theory of stochastic cooling is much more complex than this simplified description suggests; likewise, putting the technique into practice is a very tricky procedure. The particles in accelerators are generally travelling at velocities close to that of light, and the cooling system has to react fast enough, picking up and transmitting the signal and applying the kicker fields, all within the time it takes the particles to travel around part of the ring. In practice, the signal is sent across an arc of the ring so as to reach the kicker just in time to produce an electric field of the right size.

The first attempts at stochastic cooling had to await the development of electronics fast enough to make the whole task possible. In 1975, with new electronics, tests on the proton beam in the ISR showed that the vertical size of the beam could be reduced by some 2 per cent in an hour. Later, in 1976, further tests reduced the momentum spread of the beam in the ISR, and hence its horizontal size, by as much as 10 per cent an hour. Stochastic cooling worked.

Thus by 1976 there were available two methods to concentrate antiprotons into a beam suitable for a storage ring. The proposal of Cline, McIntyre and Rubbia to put antiprotons into an existing proton synchrotron was by no means an impossible dream; it was a viable proposition. The two obvious choices for such a project were the SPS at CERN and the similar machine at Fermilab, and studies began on how the scheme could be implemented at both laboratories.

The plan that evolved at Fermilab was to produce 8-GeV antiprotons from the collisions (with a target) of protons accelerated in the main ring. The antiprotons would then be *decelerated* in the Booster – a small ring normally used to take protons up to 8 GeV before injection into the main ring. At 200 MeV, the energy of the decelerated antiprotons would be low enough for them to be stored and cooled by electron cooling. CERN's plan on the other hand was somewhat simpler, for the laboratory has a basic advantage over Fermilab in the form of the PS, the 28-GeV synchrotron that started up in the late 1950s. CERN could use 26-GeV proton beams from the PS to create 3-GeV antiprotons, which could be directly stored and cooled in an additional small ring. And, for technical reasons, whereas Fermilab had to look to electron cooling, CERN could consider stochastic cooling.

In the end, Fermilab abandoned its original plans for producing antiprotons for a variety of reasons, including difficulties with electron cooling. Now, the laboratory has a scheme based on stochastic cooling, involving two new rings built one inside the other. These developments have gone on hand in hand with the laboratory's proposal, first put forward in 1975, to build a ring of superconducting magnets in the same tunnel as the 'normal' magnets of the original synchrotron. Fermilab's aim in this respect has been twofold. First, the new ring will save electrical power when running at lower energies: as the superconducting magnets offer little resistance to an electric current, less power is used up at the same current and therefore the same field. The new magnets will also be able to sustain higher fields than conventional magnets and will allow the synchrotron to reach higher energies, up to 1 TeV (1000 GeV), giving rise to the name of 'Tevatron'.

Fermilab's superconducting ring was completed in the spring of 1983, and it accelerated protons to 512 GeV at the beginning of July the same year. The superconducting ring will eventually be used as a storage ring, providing proton–antiproton collisions at a total energy of 2000 GeV. Meanwhile, however, CERN was able to forge ahead with its own plans to convert the SPS to a proton–antiproton collider.

Antiprotons at CERN

CERN was left with a golden opportunity to build the machine that could find the W and Z^0 particles – something that had become almost a Holy

Grail to experimental particle physicists. The laboratory moved fast, despite its complex structure for making decisions, which involved twelve independent-minded member states.* By June 1978 the project to convert the SPS to a proton–antiproton collider was fully approved. The machine would collide protons with antipotons at an energy of 270 GeV per beam – a total of 540 GeV, equivalent to that released when a beam of 150,000 GeV strikes a stationary target. Although the SPS can accelerate to 500 GeV, 270 GeV is the highest energy for which the magnets can be cooled when run 'flat out' as they must do in a storage ring, as opposed to a synchrotron where they are pulsed. But a total energy of 540 GeV would be enough – enough to find the W and Z^0 particles.

Even before full approval, a significant amount of work was already under way behind the scenes to plan the precise way to tackle the conversion. This work included setting up the Initial Cooling Experiment (ICE), specifically to investigate the two techniques of beam cooling. ICE used a small ring of magnets inherited from an earlier experiment to measure precisely the magnetic moment of the muon. Set up in 1977, ICE a year later showed how successful stochastic cooling could be, first with protons, and then in July 1978, when it proved that antiprotons could indeed be stored. In particular, it became clear that stochastic cooling would be better than electron cooling at the energies at which the antiprotons were to be stored before injection into the SPS, so this became CERN's choice.

The main new component for the antiproton project was the Antiproton Accumulator – the ring to cool and store the antiprotons. In addition, a new tunnel was needed to transfer antiprotons from the PS, where they would undergo some initial acceleration, to the SPS. And extra pumps were needed to improve the vacuum of the SPS, to a level of 10^{-9} torr, from its original value of 10^{-8} torr. Extra quadrupole magnets would squeeze the beams still further just prior to collision in the SPS, and new radio-frequency cavities would deal with accelerating the antiprotons as well as the protons. Of all these items, the Antiproton Accumulator was certainly the most crucial.

In the final design for the collider, the antiprotons are produced when a 26-GeV proton beam from the PS strikes a metal target. At this proton energy, the greatest yield of antiprotons is with an energy of 3.5 GeV. So the magnets beyond the target are set up to pick off 3.5-GeV antiprotons, and direct them towards the Antiproton Accumulator, part way along a transfer line built to guide protons to one ring of the ISR, and also to the SPS.

---

* In 1978: Austria, Belgium, Denmark, the Federal Republic of Germany, France, Greece, Italy, the Netherlands, Norway, Sweden, Switzerland, and the United Kingdom.

Producing antiprotons in this way is not very efficient. Each pulse of protons from the PS contains some 10^{13} particles, but yields only 2×10^7 antiprotons that reach the Antiproton Accumulator: it takes about one million protons to produce just two antiprotons. Calculations show that the beam of antiprotons in the SPS must contain at least 6×10^{11} particles if there are to be sufficient collisions to reveal interesting processes. Thus, the accumulator needs to store 30,000 (3×10^4) pulses of antiprotons from the target before it sends a beam to the SPS. This is equivalent to using more or less all the pulses from the PS during a period of twenty-four hours.

The Antiproton Accumulator is a small 'ring', in fact roughly square, comprising twelve bending magnets, and twenty-four quadrupole magnets for focusing. The magnets are unusually dumpy, having a wide aperture in the horizontal direction, both to accept as many antiprotons from the target as possible and to allow the cooling mechanism to work effectively. Once they enter the accumulator, the particles begin by circulating through the magnets at the outer edge of the vacuum chamber, as the first of the sequence of diagrams in Figure 46 illustrates.

Cooling begins almost immediately. Signals from electrodes that sense the average position of the antiprotons race the particles to a point farther round the ring where an electric field 'kicks' them by an appropriate amount, to put them nearer the optimum course round the ring. After 2 seconds, the spread in the momentum of the antiprotons has been reduced from 0.75 per cent to 0.1 per cent and at this point a shutter across the vacuum chamber is raised so that the bunch of particles can be transferred to the other side of the chamber, to form the so-called 'stack'. The shutter then closes, another pulse of antiprotons enters the ring from the target, and the 'precooling cycle' begins again.

Meanwhile, the antiprotons in the stack are being further cooled on each orbit of the ring, with corrections being made both vertically and horizontally. One hour and 1500 pulses of antiprotons later, there are as many as 3×10^{10} antiprotons in the stack, many of which are concentrated in a dense core; forty hours and 60,000 pulses of antiprotons later, there are some 10^{12} antiprotons in the stack and 6×10^{11} in the core. In this period the beam has been concentrated by a factor of 100 million – the time is ripe for extraction. Electric fields pull the dense core of antiprotons from the stack, leaving around 4×10^{11} particles behind to begin a new stack, which will be full again in twenty-four hours' time. Then another 6×10^{11} antiprotons can be extracted from the accumulator and sent en route to the SPS.

The Antiproton Accumulator is a tour de force in accelerator physics, vital to the operation of the SPS as a collider. Only two years after its construction began, it came into operation in July 1980, storing its first

Figure 46   *The principle behind the operation of CERN's antiproton accumulator. On injection into the accumulator, the antiprotons orbit at the left-hand side of the beam pipe (a), where they are precooled as they circulate for 2 seconds (b) before being transferred to the stack (c). The accumulator is then ready to receive the next pulse of antiprotons (d), which is likewise precooled and stacked 2.3 seconds later (e). After two minutes, fifty pulses have been stacked (f) and after an hour the stack contains 3×10^{10} particles from 1500 pulses (g). After forty hours, the stack is full, with a total of 10^{12} antiprotons, 6×10^{11} of which are in the dense core (h). The particles in the core are then ejected (i), leaving 4×10^{11} in the stack, and the cycle can begin again*

antiprotons. The large, 70-cm-wide vacuum chamber was carefully conceived to give a good vacuum; it is kept at 10^{-10} torr, sufficient to minimize the loss of antiprotons during their twenty-four hours (or more) of circulating the ring. With the Antiproton Accumulator working, CERN was well on the road towards the ultimate goal of high-energy proton–antiproton collisions.

To put antiprotons into the SPS involves sending the particles on a route rather like that of some giant pinball machine. At 3.5 GeV the antiprotons are too low in energy to go directly to the SPS from the accumulator. Instead, they must double back to the PS, the very machine that accelerated the protons that spawned the antiprotons (Figure 47). In the PS the antiprotons are taken up to 26 GeV, above an energy that causes some problems in the SPS, and only then are they ready to enter the SPS, grouped in six bunches, with 10^{11} particles per bunch.

On 14 February 1981, CERN's PS, the trusty 28-GeV proton synchrotron that had served the laboratory so well since 1959, became the world's first antiproton accelerator. It had successfully accelerated a beam of 5×10^8 antiprotons to 26 GeV, the energy that would be required for injection to the SPS. But the SPS was still not quite ready, having been shut down the previous June for some fairly extensive alterations in preparation for accepting antiprotons, including the construction of two subterranean caverns to house experiments at points where the beams would cross. So the world was to see the first collision of beams of antiprotons and protons not in the SPS but in the ISR, which was still producing the world's highest-energy manmade particle collisions. In the small hours of 4 April the first protons and antiprotons met in the ISR at an energy of 26 GeV per beam, or 52 GeV in total. As expected, the experiments at the intersections, including one preparing to observe collisions in the SPS, saw nothing dramatically new in comparison with proton–proton interactions at the same energy. More important, perhaps, was that the exercise showed that a high-energy storage ring would work with antiprotons. It was a rehearsal that delighted everyone involved.

In the next couple of months preparations for putting antiprotons into

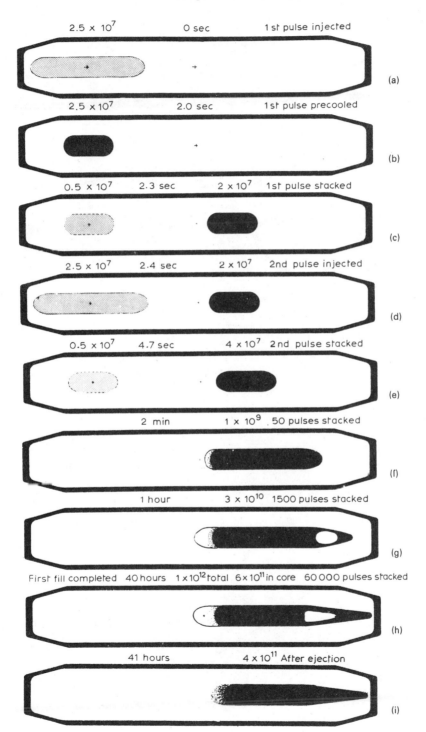

2.5×10^7 O sec   1st pulse injected   (a)

2.5×10^7 2.0 sec   1st pulse precooled   (b)

0.5×10^7 2.3 sec 2×10^7 1st pulse stacked   (c)

2.5×10^7 2.4 sec 2×10^7 2nd pulse injected   (d)

0.5×10^7 4.7 sec 4×10^7 2nd pulse stacked   (e)

2 min 1×10^9 50 pulses stacked   (f)

1 hour 3×10^{10} 1500 pulses stacked   (g)

First fill completed   40 hours 1×10^{12} total 6×10^{11} in core   60 000 pulses stacked   (h)

41 hours 4×10^{11} After ejection   (i)

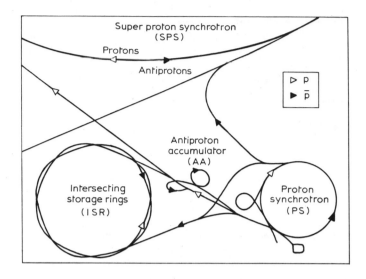

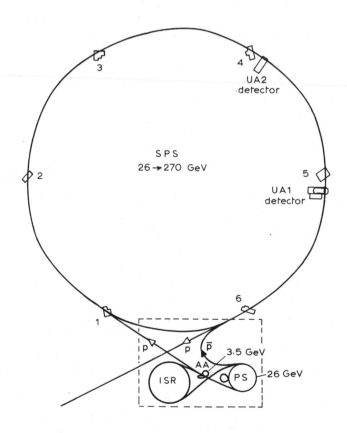

Figure 47 *CERN's antiprotons start life when a 26-GeV proton beam from the Proton Synchrotron (PS) strikes a target. Antiprotons at 3.5 GeV are selected and taken to the Antiproton Accumulator, where they are stored and cooled. Once enough antiprotons have been stored, they are returned to the PS and accelerated to 26 GeV, prior to injection into the Super Proton Synchrotron (SPS). Meanwhile, protons at 26 GeV have already been injected from the PS into the SPS, and the two beams can then be accelerated simultaneously to 270 GeV before colliding*

the SPS proceeded at full pace. At the last it seemed that the weather was conspiring to prevent the project from succeeding. CERN takes some electricity from the French national grid, and thunderstorms, common on summer nights in the Jura mountains across the border from Geneva, often succeeded in knocking out power to the SPS. But on 7 July the first antiprotons were accelerated in the SPS to 270 GeV and stored for a short while. Two days later, detectors indicated that the first proton–antiproton collisions within the SPS had occurred.

9

To Catch a Particle

A new accelerator, with access to new physics, always presents a tremendous challenge to experimenters. How can they best predict what might happen? How can they be sure of detecting what they think might happen? How can they be certain not to miss something they did not expect? CERN's proton–antiproton collider was no exception since, with ten times more energy available in collisions than ever before, except in studies of cosmic rays, it would open doors to completely new phenomena. The collisions were bound to produce more particles, with higher energies, more closely packed together, than had previous experiments. And somehow all this new physics had to be observed as cleanly and with as little ambiguity as possible.

The prize that most people hoped the collider would bring was, of course, the discovery of the missing W and Z^0 particles, the intermediate vector bosons of the electroweak theory of Glashow, Salam and Weinberg. These particles would, in theory, be formed in the fusion of quarks with antiquarks during the proton–antiproton collisions, and they would often decay by the reverse process. But quarks and antiquarks seem unable to exist alone, so those formed in the decays of the vector bosons would have to gather up additional quark–antiquark pairs from the underlying quantum fields and emerge as relatively tight showers, or 'jets', of hadrons – particles made from quarks. Given that many other kinds of interaction between proton and antiproton also produce many hadrons, picking out W and Z^0 particles through their decays to hadrons would not be easy.

Fortunately, there is another way to spot the vector bosons, for theory suggests that 20 per cent of the time a W particle should decay to produce two leptons; for the Z^0 the figure is 10 per cent. Specifically, a W^- can decay to an electron or a muon, together with its associated antineutrino to conserve quantum numbers (Figure 48a):

$$W^- \rightarrow e^- + \bar{\nu}_e; \qquad W^- \rightarrow \mu^- + \bar{\nu}_\mu.$$

Similarly a W$^+$ can decay to a positron (or μ^+) and neutrino:

$$W^+ \to e^+ + \nu_e; \qquad W^+ \to \mu^+ + \nu_\mu.$$

This is rather like the way the W particle is supposed to create an electron and antineutrino in the beta decay of a neutron. However, beta decay is a low-energy process, and the W particle involved is virtual, its mass-energy borrowed fleetingly from the quantum field in the time allowed by the uncertainty principle. In quark–antiquark fusion on the other hand, the W created is real, though it lives only a short time, a mere 10^{-25} s, before it decays to more stable particles.

Neutrinos (and antineutrinos) are notoriously difficult to detect as they interact only weakly with matter, which leaves only the charged lepton to register in the apparatus. Fortunately, the electron (or muon) and neutrino should often emerge from the decay of a W at large angles to the direction of the colliding beams, roughly perpendicular to the accelerator's beam pipe. The detection of such an electron or muon, together with the missing energy carried by the unobservable neutrino, provides a unique and relatively clear fingerprint for the W particles. In addition, the parity-violating nature of the weak force leads to an asymmetry in the production of charged leptons in the decays. The decaying W particles produce more positrons (or positive muons) than electrons (negative muons) in the direction of the outgoing antiproton beam. The detection of this asymmetry provides a further means for experimenters to satisfy themselves of what they are in fact seeing.

In the case of the Z^0 particle, the leptons provide a still clearer signal, for the Z^0 should decay into a pair of charged leptons (Figure 48b):

$$Z^0 \to e^+e^-; \qquad Z^0 \to \mu^+\mu^-$$

Here all the energy carried away by the two outgoing particles should add up to give the equivalent mass of the Z^0 particle. Again there is an asymmetry in the general directions of the leptons produced, but the effect is smaller than for the W particles and should be much more difficult to detect.

Also of crucial importance to the electroweak theory is the Higgs particle (H^0), which appears by virtue of the mechanism introduced to give the vector bosons their masses, as described in chapter 6. The Higgs boson, a scalar particle with spin 0 (as opposed to the spin 1 of the vector particles), may be created in association with a W or Z^0, in quark–antiquark annihilation or the fusion of two gluons (Figure 48d). How it decays depends on its mass, which unfortunately is not at all well predicted by theory, so our ideas as to how the Higgs particle may reveal itself are much more vague than they are for the W and Z^0 particles. It may decay to hadrons containing

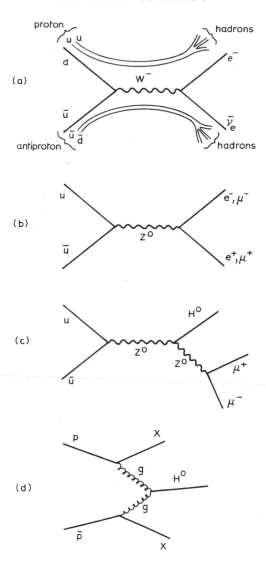

Figure 48   *The cleanest signal for the production of intermediate vector bosons in proton–antiproton collisions is through their decays to leptons. The W^-, for example, should be produced by a d quark from a proton together with a $\bar{u}$ quark from an antiproton (a); it can decay into an electron (e^-) and an antineutrino ($\bar{\nu}_e$). The other 'spectator' quarks and antiquarks interact among themselves to produce hadrons. In a similar reaction, a u quark and $\bar{u}$ antiquark can form a Z^0 (b), which can then decay to an electron–positron pair or a muon pair ($\mu^+\mu^-$). (The spectator quarks are omitted in this case to make the diagram clearer.) The collisions should also produce the Higgs particle (H^0) which can be radiated by a Z^0 (c) or produced in the fusion of two gluons (g) radiated by the colliding proton (p) and antiproton ($\bar{p}$) (d)*

charmed or bottom quarks, which in turn will produce many strange particles or many leptons through their own weak decays.

But the collider was not built only to test out the electroweak theory; it would provide access to a new, unexplored energy region under more controlled conditions than are possible in the study of cosmic rays. One area of experiment would be to see how effects observed at the Intersecting Storage Rings (ISR) developed with the tenfold increase in energy. Of particular interest were the indications from the ISR that, in some instances, particles emerging from the collision point are closely associated in two back-to-back jets, perpendicular to the direction of the colliding beams. Such behaviour was not altogether unambiguous in the data from the ISR, but it seemed to mesh well with ideas of protons containing hard constituents – the quarks – which could scatter at wide angles in collisions. Clearer observations of such behaviour at higher energies would provide crucial tests of quantum chromodynamics, the quantum-field theory of the strong force.

The higher energies might also make it possible to observe more exotic forms of matter, such as single quarks, freed by the high energy from confinement within a hadron; or monopoles, single magnetic north or south poles unassociated with the opposite pole, which should exist according to certain kinds of theory. And the collider would close in on the energies of some unusual, unexplained phenomena observed in cosmic rays. These are the so-called 'Centauro events', discovered by Japanese and Brazilian scientists working together on a detector on Mount Chacaltaya in Bolivia. In these events, a cosmic ray entering the atmosphere produces up to 100 charged particles, several times as many as expected, but virtually no neutral pions, whereas roughly half as many uncharged as charged are expected. The Centauro events had been observed at estimated incident energies of 1500 TeV, whereas proton–antiproton collisions in the SPS would be equivalent to a particle of only 155 TeV striking a stationary target. However, the so-called 'mini Centauro' events, with fewer charged particles but still too few neutrals, appeared at energies as 'low' as 250 TeV, so there was a good chance that the collider might create the same phenomenon in the laboratory for the first time.

The wide range of phenomena open to observation gives some forewarning of the complexity of the apparatus built for the collider. Detectors would have to pinpoint electrons and muons accurately to reveal the W and Z^0 particles, and make good measurements of their energy. They would have to pick up maybe as many as 100 or so charged particles, measure their energies and reveal their directions; they would have to trap neutral particles and let only the undetectable neutrinos escape. And, added to these challenges laid down by the physics, there were challenges of a more practical nature.

The detectors would have to surround the beam pipe of the SPS and thus be reliable enough to stay there for long periods once the pipe was evacuated, the protons and antiprotons stored and circulating. They would have to go underground in the accelerator's tunnel, many metres from the experimenters and their control rooms. They would have to make the most of existing technology to search for new effects in an unexplored world. And, naturally, they would have to be built within certain costs and to within a timetable, which for experiments of the necessary complexity was all too short.

From the approval of the project in 1978, only three years were anticipated before the first collisions in 1981. CERN was determined to beat the world, which in practice meant to be at the new energies before the USA. As it happened, the American challenge proved to be weakened both by financial constraints and technical difficulties in building a superconducting proton–proton collider. However, once begun, the impetus remained with the collider project at CERN and the race became one for the experimenters to ensure that their detectors were ready when the first proton–antiproton collisions occurred in the SPS.

A general purpose detector

The initial plan at CERN was to have only one experiment to observe proton–antiproton collisions, and work began in 1977 on the design of a detector that would do everything and, it was hoped, miss nothing. A key figure in the work and spokesman for the experiment was Carlo Rubbia, who was also a driving force behind CERN's decision to proceed with the collider. By January 1978 he had formed the basis of a collaboration of eight institutions from Europe and one from the USA, which together presented a proposal for the all-purpose detector that became known as UA1 (for 'underground area', experiment 1). Since then, three more institutions have joined the collaboration to make it the biggest CERN had ever known; the team of physicists and engineers now totals in the region of 135.*

The basic aim underlying the design of the apparatus was both to record as much information as possible about the tracks of the particles produced, and to measure completely the energy they carry, over as

---

* The UA1 experiment was originally proposed by fifty-two physicists from Aachen, Annecy, Birmingham, CERN, College de France (Paris), Queen Mary College (London), Riverside (California), Rutherford Appleton Laboratory and Saclay. Since then the collaboration has been joined by teams from Helsinki, Rome and Vienna, along with occasional visitors from other institutions.

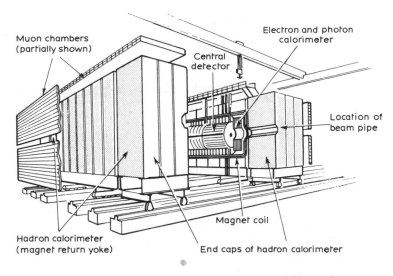

Figure 49 *The UA1 apparatus is like a huge 'jam roll' of different detectors wrapped round the beam pipe in which the proton–antiproton collisions occur. Each layer of detectors is designed to pick up a particular variety of particle, thus keeping a 'balance sheet' of as many of the products of the collisions as possible*

much of the space immediately surrounding the beam pipe as possible. The result is a huge layered concentric 'jam roll' of detectors wrapped round the beam pipe, covering essentially all the azimuthal angle (in the plane perpendicular to the pipe), and from 5° to 175° in angle along the beam pipe (Figure 49). Further 'forward' detectors, placed outside the main apparatus, close the coverage down to within 0.2° of the beams themselves.

At the heart of UA1 lies the cylindrical central detector, 5.8 m long and 2.3 m in diameter, which provides an image of the tracks of charged particles spilling out from collisions. The image is not directly visual, but is reconstructed electronically from information contained in electrical pulses from a network of over 6000 'sense' wires crossing the six sections of the detector. Three sections each nearly 2 m long hug the beam pipe from above, three from below (Figure 50). In the central two sections, horizontal wires lie in planes aligned vertically with respect to the beam pipe; in the end sections, four in all, the planes of wires lie horizontally. Each section of the detector is filled with a mixture of ethane and argon gas and subject to an electric field set up by holding alternating planes of wires at a potential of −30 kilovolts. The field is thus horizontal in the central sections and vertical in the outer sections. The whole detector sits within a huge dipole magnet in such a way that the magnetic field is in

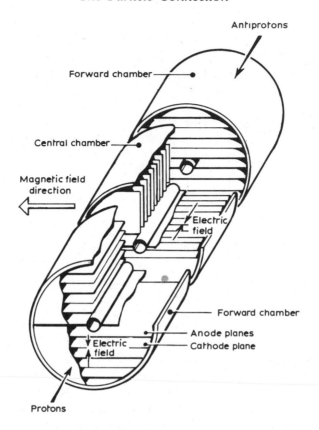

Figure 50   *The central detector of UA1 comprises six sections, each with planes of wires strung horizontally. In the 'forward' chambers the wires form horizontal planes, so the electric field between them is vertical. In the central chambers, the wire planes are vertical, and the electric field is horizontal. However, in all the chambers the electric field is perpendicular to the magnetic field*

a horizontal direction, at 90° to the beam pipe; this means that in all six sections the magnetic and electric fields are mutually perpendicular.

Charged particles emanating from a collision at the centre of the detector ionize the gas as they travel outwards. Electrons released by the ionization drift towards the planes of wires at zero voltage (which are positive relative to the high-voltage planes). The pattern of charges the wires pick up reflects the path of each charged particle, which will be curved by the magnetic field according to momentum, with particles of lower momentum being bent the most. By measuring the time it takes the electrons to drift to the wires – a few microseconds at most – as well as the locations along the wires where the charge is picked up, the chamber records a three-dimensional picture of the tracks of the charged

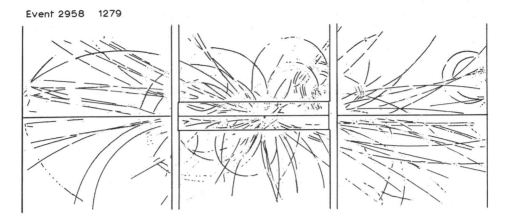

Figure 51 *A computer graphic display reveals the complex pattern of tracks left by particles as they fly through UA1's central 'image chamber' from a proton–antiproton collision at the centre point. (From: G. Arnison* et al., Physics Letters, *vol. 122B (1983), p. 103)*

particles, which can later be reconstructed on computer graphic displays (Figure 51).

The central detector should reveal all charged particles – mainly protons, pions, electrons and muons. Part of the function of the remaining detectors is to sort out which of these is which. The next layer of the jam roll concentrates on electrons, together with any photons produced in the collisions or in the decays of neutral pions, which will not have registered in the central detector. This layer is the electromagnetic calorimeter, designed to measure the energy of photons and electrons, and to identify electrons among a host of charged hadrons. It comprises forty-eight vertical gondola-shaped structures, with an inside radius of 1.36 m; twenty-four are wrapped vertically round each side of the central drift chamber. Each gondola is a sandwich of many layers of lead and plastic scintillator, a material that emits tiny flashes of light when a charged particle passes through and excites atoms within the plastic.

How does the lead–scintillator sandwich give a measure of the energy of the electrons and photons? The principle is that in the lead, with its high density of charged atomic nuclei, electrons will tend to radiate photons; conversely, photons will tend to produce pairs of electrons and positrons, provided, of course, they have sufficient energy; and the electrons and positrons can in turn radiate new photons and so on. Thus, in traversing the lead, an electromagnetic shower of photons, electrons and positrons develops, each individual particle having successively less energy as the number of particles sharing the original amount of energy

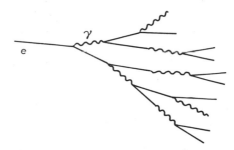

Figure 52   *An electron, radiating a photon in the electric field of an atomic nucleus, initiates an electromagnetic shower. Photons with enough energy produce electron–positron pairs; electrons and positrons can radiate photons and can also annihilate, creating photons. The shower peters out when the individual particles have too little energy for these processes to continue to occur*

increases (Figure 52). The shower effectively 'runs out of steam' and comes to a halt when the energy becomes too small for the radiation and pair-production processes to occur.

The idea of the lead–scintillator sandwich is to sample that shower at many points during its development. The amount of light from each sheet of scintillator gives a measure of the energy deposited at successive stages in the shower. This light from the scintillator is allowed to propagate, along plastic light guides, to photosensitive tubes that convert the light into an electrical signal proportional to the number of photons. In this way the detector builds up a picture of the shower's development and the energy contained in the initial particle. Thus the lead–scintillator gondolas trap electrons and photons from the proton–antiproton collisions, but the heavier muons and the hadrons pass through, to be caught in other layers.

Outside the gondolas lies the aluminium coil of the electromagnet, supplying the field to analyse the momentum of the charged particles. The magnetic field of 0.7 teslas uniformly fills a region $7 \times 3.4 \times 3.3$ m^3 around the central detector and gondolas, which nestle between the two vertical halves of the coil. Outside the coil the field is contained in an iron 'return yoke', which is divided into sheets interleaved with plastic scintillator. The whole yoke is thus an intricate iron–scintillator sandwich, divided into sixteen vertical sections shaped like rectangular 'C's which surround the coils, eight on either side. The sheets of scintillator convert the iron yoke into another calorimeter, this time to study the showers created by hadrons in nuclear collisions within the iron. The principle is the same as before, with light guides taking signals to phototubes that output small electrical pulses proportional to the amount of energy deposited. By splitting each C-shaped section into twelve portions, the hadronic calorimeter gives a measure of the direction in which energy flows out from the collisions, as well as the amount. The yoke contains some 830 tonnes of iron, in sheets 5 cm thick and 80×90 cm^2, interleaved with around 2000 m^2 of plastic scintillator, viewed by a total of 768 phototubes.

The only particles, other than neutrinos, to penetrate beyond the colossal hadron calorimeter should be muons, which do not interact strongly and so pass easily through the iron. To detect the muons, which unlike the neutrinos are electrically charged, the outer layer of UA1 comprises two sets of flat detectors 60 cm apart. Each is built like a raft from four layers of 'drift tubes'. These act in a similar way to the central chamber in that charged particles crossing the tubes ionize the gas they contain, and the electrons released drift in an electric field to (positive) wires along the centre of each tube, where they are collected. Muons will produce tracks in these chambers that should connect back to the tracks they made in the central detector, thus singling them out from electrons, which would stop in the gondolas, and hadrons, which become absorbed in the iron.

To complete the main body of the UA1 detector are 'end caps', which enclose the apparatus with layers of lead–scintillator and iron–scintillator, matching up with the gondolas and C-pieces of the magnet yoke. A further 620 tonnes of iron and 24 tonnes of lead make up the end caps, bringing the total weight of the apparatus to some 2000 tonnes. Additional planes of drift tubes intercept muons emerging through the end caps.

The main bulk of the apparatus covers angles between 5° and 175° along the beam pipe. On either side, compensating magnets help to counteract the effect of the dipole field on the circulating beams, and these are converted into calorimeters just as the main magnet is, to detect particles emerging at angles closer to the beam. In addition, there are detectors to pick up showers at still smaller angles, and drift chambers to increase the field of the image from the central detector. Together these 'forward' detectors narrow down coverage to within 0.2° of the beam. Additional small drift chambers, in fact inside the beam pipe, cut off from the vacuum in which the beams travel by only a thin window, monitor the intensity of proton and antiproton beams at points 22 m on either side of the collision point.

The whole apparatus was brought together below ground in a huge area especially excavated around the collision point. The area is basically two cylindrical halls, roofed over and connected (Figure 53). One hall is centred on the collision point; the other is a 'garage' where the apparatus was assembled on rails. The 2000-tonne colossus can be moved on the

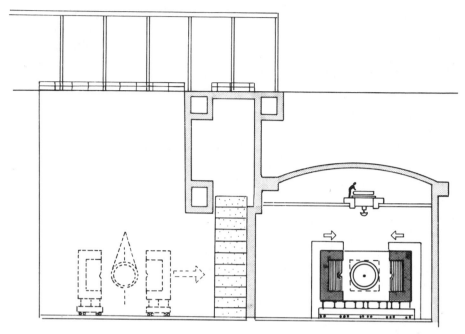

Figure 53 *The UA1 apparatus was put together in the underground 'garage' (left) from where it can be moved on rails to its position in the SPS ring (right). When the synchrotron is in operation, a wall of concrete blocks shields the garage area from radiation in the ring*

rails from its resting position in the garage to its working position in the SPS ring, and back again when the SPS is used for other experiments. Concrete blocks are installed between the halls when the SPS is operating to shield the garage from radiation.

The thousands of signals produced each time a collision is recorded are passed to a mobile electronics hut, where processors make fast decisions. From there signals go up to the control room at ground level, 100 metres away. Bunches of protons and antiprotons meet at the heart of the detector every 8 microseconds when the accelerator is working, and sophisticated electronic 'trigger processors' make the initial decision as to whether a collision is worth recording or not: it takes some 35 ms to record all the information from the whole apparatus, in which time several hundred beam-crossings must be ignored. Therefore it is of paramount importance to ensure that the collisions recorded are the most interesting ones. One trigger processor, containing thousands of electronic circuits, looks at the patterns of energy flow in the two calorimeters and makes its decisions on the basis of this information, according to preprogrammed instructions that 'tell' the processor the characteristics of desirable 'events'. The other processor looks particularly for muons coming directly from the proton–antiproton collisions, rather than from decays of other particles within the apparatus. In addition, a data-processing system reduces around 2 million bytes, or computer 'words', of information per event to about 100,000 bytes, which are then stored on magnetic tape for later analysis.

A second experiment, UA3, lies at the heart of UA1. This consists of detectors to search for monopoles, particles carrying single magnetic poles, which should exist according to a number of theories. Indeed, in 1931, Paul Dirac showed that quantum theory predicts the existence of monopoles of strength $hc/2e$, where h is Planck's constant, c is the velocity of light, and e is the electron's charge. As yet no convincing evidence for monopoles has been found, but as their masses are generally not predicted, it is always reasonable to explore a new energy region in search of them.

The single magnetic charge should lead to an unusually high amount of ionization when a monopole passes through a material. The technique exploited in UA3, by four physicists from CERN and Annecy, is to use a special plastic, called Kapton, in which monopoles will leave heavily ionized tracks, while the usual lightly ionizing particles produced in the proton–antiproton collisions will leave no tracks at all. The tracks become visible in the plastic when it is immersed for a few hours in sodium chlorate. The acid etches away the ionized tracks more rapidly than it does the normal material, and so the tracks appear as tubular holes along the paths of the particles that produced them. Monopoles should produce tracks that are not only well ionized but which also extend much farther than those of heavily ionizing nuclear fragments produced in the collisions, which will penetrate only a millionth of a metre (1 micrometre) or so of the plastic. The UA3 team introduced several layers of Kapton, 75 micrometres thick, inside and outside the beam pipe through UA1, round the central image chamber of UA1 and between the three sections along the beam pipe. After that it was a matter of waiting.

The second intersection

CERN originally agreed to excavate an area for experiments around one point where the proton and antiproton beams were to cross in the SPS. The project aroused so much interest, however, that by December 1978 CERN had agreed to open a second area, and had selected an experiment that became code-named UA2. Built by a collaboration of scientists from six institutions,* UA2 was slightly more modest in its aims than UA1.

---

* The UA2 collaboration is between around fifty physicists from Bern, CERN, Copenhagen, Orsay, Pavia and Saclay.

Nevertheless, it is by no means of modest size and its construction presented similar technical challenges. The scientists behind UA2 were equally interested in the possibility of finding W and Z^0 particles, but decided to concentrate on the decays to electrons, and on the 'forward–backward' asymmetry of the decays with respect to the beam direction. They were also particularly interested in making good measurements at large angles, around 90° to the beam pipe, so as to study the production of transverse jets of particles at the new high energies.

The central part of the detector consists of a number of wire chambers, designed to locate the tracks of charged particles as they leave ionized trails in gas. Unlike UA1, this experiment has no magnetic field in this central region, so the particle tracks are straight, and the collision point can be determined very precisely. A cylinder of tungsten, about 0.5 cm thick, surrounds the central detector, immediately followed by another cylindrical wire chamber. The tungsten initiates electromagnetic showers when photons or electrons pass through, and these produce telltale signals in the outer wire chamber.

Outside the central chamber lies a large calorimeter: a detector to pick up all the energy from electrons and photons and from hadrons (Figure 54). The calorimeter is divided into many sections to provide as detailed a picture as possible of the flow of energy from a collision. It is basically spherical and is divided rather like an orange into twenty-four horizontal segments surrounding the beam pipe. Each segment, or wedge, is further divided from one end to the other into ten sections, so that the whole consists of 240 cells. Each cell, 1.13 m long, contains elements for absorbing first electromagnetic and then hadronic energy.

The portion of each cell nearest the collision point is a multiple sandwich of lead and scintillator, which samples the energy deposited by electromagnetic showers that form in the lead. Behind this, farther from the collision, lie two iron–scintillator sandwiches, to monitor the energy of hadron showers. Light guides take the light emitted in the scintillators to phototubes arranged outside the 'orange', where they bristle like the spines on a sea urchin.

The segmented calorimeter covers an angle of between 40° and 140° along the beam pipe; particles emerging closer to the beam pipe, to within 20°, are intercepted by different detectors. On either side of the central calorimeter, toroidal (doughnut-shaped) magnet coils encircle the beam pipe. They are followed by drift chambers and other wire chambers to record the trajectories of charged particles, and further lead–scintillator detectors to collect energy from electromagnetic showers initiated by high-energy electrons.

The collision point monitored by UA2 is at one of the deepest parts of the SPS tunnel, 63 m below ground. There the apparatus was assem-

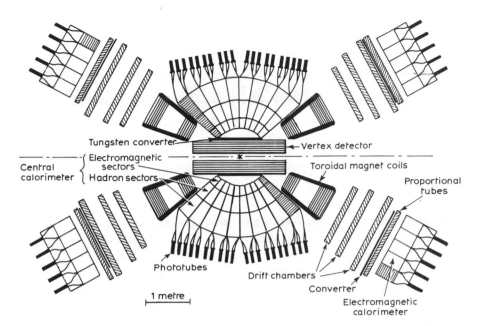

Figure 54 *A schematic cross-section through the UA2 apparatus. The vertex detectors consist of wire chambers designed to track charged particles in a region with no magnetic field. The central calorimeter, segmented like an orange, surrounds the vertex detector, except at 'forward' angles, within 37.5° of the beam pipe. This region is covered down to 20° by a symmetric arrangement of wire chambers and electromagnetic calorimeters encircling the beam pipe beyond toroidal magnet coils*

bled in a huge underground cavern, excavated around the beam pipe. The 200-tonne detector glides into place in the SPS ring on air cushions, and UA2 can observe proton–antiproton collisions at the same time as UA1.

UA2 shares its underground location with another experiment, UA4.* This apparatus is designed to measure the elastic scattering of protons and antiprotons, as well as the total cross-sections – in other words the overall probability for proton–antiproton interactions of all kinds. Such information is important if the collider is to reveal how trends in particle interactions develop in the new high-energy region. Because the protons and antiprotons have very high energies, elastic scatters, in which the proton and antiproton behave like billiard balls and retain their identities, tend to be in the direction in which the particles were originally travelling – the 'forward' direction. This means that elastically scattered particles

* The UA4 collaboration is between twenty or so physicists from Amsterdam, CERN, Genoa, Naples and Pisa.

will emerge from the collision point close to the majority of beam particles which do not interact at all. To observe such particles, the UA4 team, uses tiny drift chambers mounted within the beam pipe, some 40 m distant from the collision point in each direction. A steel plate, a mere 0.1 mm thick, isolates these chambers from the vacuum necessary for the proton and antiproton beams to survive. Together, the chambers pick up particles scattered by as little as one hundredth of a degree.

A larger 'telescope' of drift chambers and sheets of scintillator, mounted either side of the UA2 apparatus along the beam, monitors inelastic scattering, in which particle identities change and new particles can be produced. These detectors cover a larger angle, from 10° down to 0.5° of the beam pipe. Information from this telescope can be extrapolated back to the actual collision point and provides data on the total number of inelastic collisions. These results together with the data from the elastic-scattering detectors give a measure of the total cross-section – or probability – for proton–antiproton reactions.

But neither UA4 nor UA2 was the first experiment to observe proton–antiproton collisions at the second collision point. The first events, immediately after the accelerator team had successfully made protons and antiprotons collide, were observed by a relatively simple visual detector, designed not to search for W and Z^0 particles but to gain an immediate impression of the new energy range, even with relatively few collisions available. This experiment, code-named UA5,* consists basically of two large boxes, each 6 m long, 1.2 m wide and 0.72 m high, containing a mixture of helium and neon gas. The boxes fit above and below the beam pipe, 9 cm apart. Charged particles from the collisions ionize the gas, which is almost immediately – within about one millionth of a second – subjected to a high-voltage pulse. The high electric field created makes the ionized gas break down and small luminous streamers, some 5 mm long, form along the ionized trails left behind by the charged particles.

Three cameras view each chamber: two provide independent images which can be used to create a stereo view of the tracks within the chamber; the other has a wider field of view and provides a back-up. The cameras are equipped with image intensifiers with a gain of 2000, to produce as clear a picture as possible of the faint pink streamers, so that adjacent tracks as close as 2 mm can easily be resolved when the developed film is examined. Banks of scintillator sheets, located around the beam pipe on either side of the streamer chambers, provide a signal to trigger the high-voltage pulse when an interesting collision has occurred.

UA5 had a useful rehearsal at CERN's Intersecting Storage Rings,

---

* The UA5 collaboration is between around forty-five physicists from Bonn, Brussels, Cambridge, CERN and Stockholm.

when antiprotons were stored in one ring in April 1981, and CERN witnessed its first collisions between beams of protons and antiprotons. The data revealed nothing dramatic at the total collision energy of 63 GeV, a region that has been thoroughly explored with proton–proton collisions. But UA5 was able to produce pictures of the high-energy reactions and give some taste of things to come. A couple of months later the apparatus was installed in the ring of the Super Proton Synchrotron, ready and waiting for the first collisions and a glimpse of a new range of energies.

10

The Taste of Success

In July 1981 CERN, the showpiece of European cooperation, became the envy of the world when particle detectors showed that the beams of protons and antiprotons in the SPS had actually collided for the first time at a total energy of 540 GeV. But the effort to find the W and Z^0 particles was far from over, neither for the engineers and physicists in charge of the accelerators, nor for the experimenters who were still frantically preparing to observe the collisions, which had finally been produced a month or so ahead of schedule. There were many finishing touches to the apparatus to be made but, perhaps more important to the whole project, the machine physicists had still a long way to go to produce a sufficient number of collisions.

A total energy of 540 GeV may seem high compared with the predicted masses of the W and Z^0 particles, of only 80–90 GeV, but remember that the colliding proton and antiproton each contain three constituents – the quarks and antiquarks – all of which must share the total energy. The vector bosons should be formed in the fusion of but one quark with an antiquark, so the energy available for their production is significantly less than the total energy of 540 GeV. This in turn means that the vector bosons should be very rare, as the fused quark–antiquark pairs will not always have just the right energy. Thus the total number of collisions was all-important.

A measure of the number of collisions between two beams of particles is given by a parameter known as the luminosity. This depends on the number of particles in each beam, the number of bunches of particles, the time between the bunches, and the cross-sectional area of the beam, and it gives a measure of the number of particles passing through the collision region each second. CERN hoped for a luminosity in the collider of around 10^{30} per square centimetre per second (10^{30} cm^{-2}s^{-1}). Of course, there are nowhere near so many particles in the beams, and the effective cross-section of the colliding zone is much less than a square centimetre; rather, the figure reflects the required 'density' of particles in the collisions. Calculations based on electroweak theory showed that a luminosity of 10^{30} cm^{-2}s^{-1} would be just high enough to produce perhaps some ten W particles and one Z^0 in the apparatus of UA1 or UA2 each day. To reach this luminosity in the collider, the Antiproton Accumulator would have to be able to provide as many as 6×10^{11} particles in each delivery to the SPS; and the SPS would have to keep these particle beams well concentrated during the ensuing acceleration to 270 GeV.

The collider started to work in the summer of 1981 with a luminosity of 10^{26} cm^{-2}s^{-1}, a factor 10,000 down on the desired value, and a figure that would have to improve if the experiments were to discover W and Z^0 particles without having to accumulate data over a period 10,000 times longer than planned! It was early days, of course, and the collider's performance did improve; towards the end of 1981 it reached a luminosity of 5×10^{27} cm^{-2}s^{-1}. This enabled all the experiments to collect data and some interesting results began to emerge, although the luminosity was still too low for searches for W and Z^0 particles to begin in earnest.

The SPS was due to run as a collider again in the spring of 1982, but misfortune beset the UA1 experiment. First, the central drift chambers were contaminated by dirty compressed air which was used by accident to cool the SPS vacuum pipe, and these delicate pieces of equipment had to be cleaned carefully. Later, a large-diameter waterpipe burst and flooded the underground cavern to a depth of 2 m or so. Meanwhile, CERN pressed on with its programme for other kinds of particle physics, and it was not until the autumn of 1982 that the SPS worked once more as a collider. With the aid of special magnets ('low beta' quadrupoles), which squeeze the beams down to a very small size just before they collide, the accelerator's performance continued to improve, passing a luminosity of 10^{28} cm^{-2}s^{-1} for the first time in mid-October. Further improvements came through circulating more bunches of protons and antiprotons – three of each type – and through increasing the supply of antiprotons from the accumulator. By the end of November, a few days before the period of machine running came to an end, the luminosity reached a record 5×10^{28} cm^{-2}s^{-1}. By now both UA1 and UA2 had collected enough data to reveal at least the presence of the W particle, if it were indeed to be produced in the way that electroweak theory suggests: a luminosity of 10^{28} cm^{-2}s^{-1} should give on average one W particle every ten days.

Members of both teams were hard at work from the beginning of this session of data taking, analysing the information collected for evidence for the decay of a W particle. The task is by no means easy, and the cliché of 'looking for a needle in a haystack' is quite appropriate. Each collision can produce many particles, 100 or more in some cases; the complexity of some events is well revealed by an image of the tracks of charged particles through the central drift chamber in UA1 (Figure 51). To find a W particle, the teams were looking for events characterized by a single energetic electron together with missing energy, carried away by the undetectable neutrino, as described in chapter 9. However, the W particle is formed by the fusion of only one quark from the proton and one antiquark from the antiproton; there remain two more quarks and two antiquarks which must materialize as hadrons – perhaps as several tens of hadrons at the high energies of the collider. So the swarm of 'hay' in UA1's central detector may well contain the 'needle' of a fast electron, the signature of the demise of a W particle. And although both UA1 and UA2 might selectively record only those events more likely to contain the decay of a W particle, the data would necessarily contain a great deal of 'hay' to sift through.

Both teams had therefore to develop sophisticated computer programs to analyse the data thoroughly 'offline', after it had been recorded on magnetic tapes. These programs are designed to consider each event – the complete record of an accepted collision – and apply a series of criteria, each tighter than the last, which together describe only the effect expected of a W decay and nothing else. As the net of constraints tightens, more events are rejected until only those that appear like the signature expected for the W's decay are left – if any!

Search for the W

For UA1 the net begins to close with the decision as to whether an event is to be recorded. The 'trigger processor' demands that there is a signal in two adjacent cells of the electromagnetic calorimeter, which corresponds to an energy deposited along a direction transverse to the beam pipe of at least 10 GeV; this step singles out those events that may contain energetic electrons. (Other criteria characterize other kinds of event that are equally interesting, and which are also recorded.) In the thirty days of data taking in November and December 1982, the experiment recorded 140,000 such 'electron triggers', out of a total of nearly one million events, selected from some thousand million collisions observed. The analysis programs had to sift through 1000 magnetic tapes filled with data, some physicists concentrating on this aspect of the experiment, while others kept watch on the apparatus as it steadily recorded yet more collisions.

The analysis begins by selecting those events with a transverse energy greater than 15 GeV in the gondola-shaped electromagnetic detectors. This reduced the number of events obtained in the autumn of 1982 to 28,000 – a factor of 5 down on the original number. Then the programs turn to study the information from the central image chamber, and demand that there should be clean track emerging from the collision point, with high momentum (greater than 7 GeV) transverse to the beam pipe. By now there were only one fiftieth of the events left in the net, or 2125 from the initial sample.

The next step is to make certain that the track found in the central detector is from an isolated particle, unassociated with a jet of particles formed by some process other than the decay of a W; and to ensure that this track connects with cells in the electromagnetic calorimeter that have absorbed a large amount of transverse energy. By this stage there were only 167 events left.

This sample is reduced still further when the program demands that very little energy (less than 600 MeV) appears in the appropriate section of the hadron calorimeter, thus helping to confirm that the selected track is indeed due to an electron, rather than a light, high-energy hadron, such as a pion. A further check is possible by comparing the momentum of the particle as measured by the curvature of the track in the magnetic field in the central detector and the energy deposited in the electromagnetic calorimeter. This left thirty-nine events, and at this stage the physicists in a sense took over from the computer. They studied the information for each individual event, displayed by the computer in an 'interactive' mode, which allows the physicist to 'talk' to the computer and look at the event from different angles and so on.

Out of the thirty-nine events that had survived the computer analysis of the original 140,000 triggers, thirty-four proved on closer, human, inspection not to contain isolated electrons. In some cases, the selected track emerged opposite a jet of particles, produced no doubt in the same process; in others, the track was actually part of a jet, but had managed to escape earlier traps; lastly, some tracks were clearly due to one partner of an electron–positron pair produced by a photon, perhaps in the wall of the beam pipe. The remaining five events, however, looked good; moreover, there appeared to be energy missing in the direction almost back-to-back with the isolated track, just as expected for W decay (Figure 55). On the other hand, the discarded events with a single jet did not seem to have much missing energy. A similar analysis, starting again with the electron triggers, but this time picking up energy deposited in the petal-shaped electromagnetic calorimeters of the end caps rather than the gondolas of the central part of the apparatus, revealed one more such event, with an isolated electron-like track with no accompanying jet.

To make certain of themselves, the UA1 team went back to the 2125 events with a clean high-transverse-momentum track in the central detector, and re-analysed them, this time concentrating on criteria to pinpoint missing transverse energy that would indicate the escape of a neutrino. Once again, after a final detailed examination of events displayed

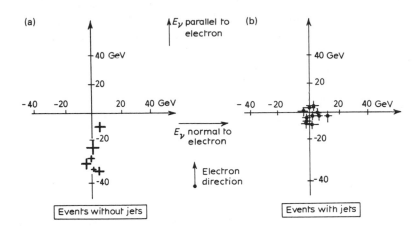

Figure 55 *The last stages in UA1's search for W decays. Events with a single high-energy electron and no jet of particles opposite have energy missing in the direction opposite to the electron's path (a). Similar events, but with a jet, show no missing energy (b). If the missing energy is carried away by an undetected neutrino, then events in plot (a) are consistent with W particle decays. (From: G. Arnison et al., Physics Letters, vol. 122B (1983), p. 103)*

on the computer, they found they were left with the same five events that the previous analysis (looking at energy deposited in the gondolas) had isolated. In other words, there were five events that looked good, with all the hallmarks of the decay of a W particle (Figure 56).* Four of the events contained electrons and therefore corresponded to the decay of a W⁻; the fifth contained a positron and thus matched the decay of a W⁺. (The sixth event discovered, with energy deposited in one of the end caps, had one or two features that made the team decide to reject it in their final analysis.)

So out of one thousand million proton–antiproton collisions, the UA1 team had found five occasions on which a high-energy electron-like particle emerged at a wide angle to the beam, with energy missing in the opposite direction, consistent with the departure of a neutrino. But what more could they say about the events? By adding the transverse energy of the electron to the missing energy they could obtain a lower estimate of the mass-energy of the particle that might have produced the electron and neutrino. A direct measurement of the full mass is precluded by the fact that UA1 cannot measure *all* the energy emerging from the collisions; as some particles are always 'lost' along the beam pipes, the apparatus has an accurate measure

---

* Figure 51 also contains the decay of a W particle. The lowest track in the lower right-hand side of the central detector is the track of the high-energy electron from the decay.

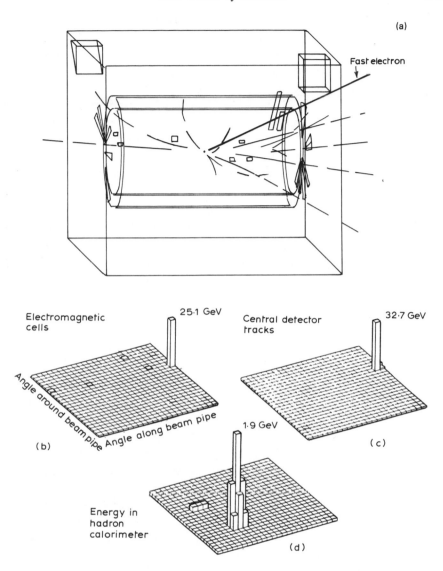

Figure 56 *A relatively clean collision (a), producing a W particle which decays into a fast electron (arrowed track) and an undetectable neutrino. This computer graphic display shows tracks in the central image chamber and those cells of the surrounding calorimeters that have 'fired'. The large rectangle outlines the extent of the hadron calorimeter. The dotted lines extrapolate the tracks to detectors outside the main calorimeter. The 'Lego' plots for the same event show the energy deposited across the angular range of the apparatus in (b) the electromagnetic calorimeter, (c) the central detector, and (d) the hadron calorimeter (on a much expanded scale). The electron produces a sharp spike in plots (b) and (c), but leaves no energy in the hadron calorimeter (d). (From: G. Arnison et al.,* Physics Letters, *vol. 122B (1983), p. 103)*

only of energy transverse to the beam direction. However, the total transverse energy showed that the parent particle must have a mass at least of 73 GeV. Going one step farther, the team could assume that the decaying particle was indeed a W, and compare with theory the angle of the emitted electron, and the energy both measured and missing. The result is a value of 81 ± 5 GeV, well in agreement with the predictions of the electroweak theory. Moreover, the number of events observed, vis-à-vis the number of collisions, also agrees well with theoretical estimates.

How did the UA2 team fare with the data collected over the same period? Like UA1, UA2 had observed some 10^9 proton–antiproton collisions of which they selected one million to record. These were events in which more than 8 GeV of energy had been deposited in a 2×2 group of adjacent cells in the segmented central calorimeter. And as with UA1, complex computer programs tailored to the specific apparatus had to sift through the data looking for the proverbial needles.

UA2's analysis begins with a search for isolated electrons, and the first test is to check the energy deposited in the calorimeter. The program demands that less than 10 per cent of the energy lies immediately outside the 2×2 cluster of cells. Moreover, tests in an electron beam at CERN had shown that high-energy electrons would lose over 90 per cent of their energy in the electromagnetic portion of the calorimeter, so the program also requires that less than 10 per cent of the energy associated with a track should be deposited in the hadronic cells. The transverse energy is calculated for the remaining events, and only those where it is greater than 15 GeV are retained. This left 363 events for the next stage of the analysis, in which each event is fully reconstructed making use of all the information from the central wire chambers.

Once the tracks of the particles have been calculated, the analysis program demands that one and only one track points to the cluster of cells in the calorimeter where the electromagnetic energy has been deposited, and this reduced the number of events to ninety-six. The remaining events are then scrutinized to study the behaviour in the chamber outside the layer of tungsten wrapped round the inner track detectors. An electron should initiate a shower in the tungsten, which will then produce a characteristic cluster of pulses in the chamber beyond. Thirty-five events had a pulse characteristic of electrons. For an electron from W decay one would expect no other large pulse in this chamber within 10° of the selected pulse; only ten events satisfied this criterion.

The last check is on the distribution of the energy deposited in the electromagnetic cells, to see if this agrees with that expected for an electron. Only three of the remaining ten events passed this test; these same three events were selected by physicists examining individually the whole sample of 363 events on computer displays.

The UA2 team also scanned through the recorded events for likely candidates with single electrons in the 'forward' detectors, the chambers and calorimeters beyond the toroidal magnets on either side of the central calorimeter. In this case, analysis revealed 761 events with more than 15 GeV of transverse energy deposited in two adjacent electromagnetic cells. The next step is to use the information from the central track detector, to ensure that a single track lines up with the appropriate cells in the calorimeter; that there is agreement between the measured values of energy and momentum (known from the bending in the fields of the toroidal magnets); and that the track originates from the collision, rather than from the conversion of a photon in the inner layers of the central wire chambers. Four events survived this detailed examination, but one of these was rejected because too much energy was deposited in the hadronic section of the calorimeter. So a total of six events with high transverse energy in either the central or forward parts of the UA2 apparatus appeared to contain an isolated electron (Figure 57).

The next test is to search for the implied presence of a neutrino by assessing the missing momentum, calculated for each event from all the information gathered in the various detectors. Four of the surviving events had missing momentum that was more than 80 per cent of the energy associated with the suspected electron, just as expected for the decay of a W particle. The other two events, both in the forward part of the apparatus, were not consistent with electrons resulting from W decay.

The remaining four events are entirely consistent with the decay of a W particle into an electron and a neutrino, both with high energy transverse to the beam pipe. Supposing that these events are indeed due to the decays of Ws, the UA2 team can use the theoretical expectations for the kinematics of the decay to calculate a likely mass. The result is 80^{+10}_{-6} GeV, in line with the predictions of electroweak theory, and in agreement with the value found by UA1.

The period of proton–antiproton collisions in the SPS came to an end on 6 December 1982, when the machine closed down for maintenance. But the signs of the likely candidates for W decays were already emerging from the huge amounts of data from UA1 and UA2, so intensive and efficient had been the analysis, which had proceeded along with the data collection, following only a few days behind. By the time a workshop on colliders was held in Rome on 13 January 1983, both teams had clear evidence of isolated high-energy electrons emerging from a haze of low-energy particles. On 20 and 21 January respectively, UA1 and UA2 presented their results to packed audiences in seminars at CERN; and by 25 January the news was definitely out when CERN informed the

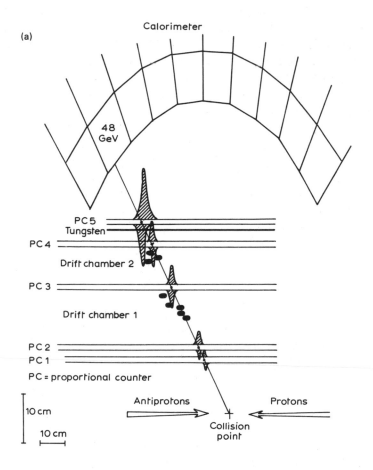

(a)

Calorimeter

48 GeV

PC5
Tungsten

PC4

Drift chamber 2

PC3

Drift chamber 1

PC2
PC1

PC = proportional counter

10 cm

10 cm

Antiprotons   Protons

Collision point

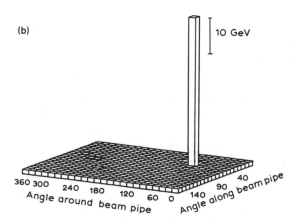

(b)

10 GeV

360 300 240 180 120 60 0 140 90 40
Angle around beam pipe   Angle along beam pipe

Figure 57 *A plot illustrating the energy deposited by a single fast electron in the central vertex detector and calorimeter of the UA2 apparatus (a). The sizes of the signals indicate the energy deposited in the proportional counters. The distribution of the energy across the angles covered by the apparatus (b) reveals that the energy carried by the electron far outweighs that of any other particle. (From: M. Banner* et al., Physics Letters, *vol. 122B (1983), p. 476)*

world's press, rather tentatively, of its latest discovery. At this stage no one officially claimed that the W particle had at last been found; rather that *evidence consistent with* the production and decay of a W particle had been observed. But, a jubilant Carlo Rubbia was less cautious only a day later at a conference in New York, when he titled his talk, 'Observations of $W^{\pm} \rightarrow e^{\pm} + v$'.

So, by the beginning of 1983, it seemed that CERN's proton–antiproton collider had attained at least one of the anticipated goals, with a total of nine likely W particles, even though it had so far fallen short of the desired luminosity. There was a chance, if the electroweak theory was correct in detail, that a Z^0 particle could also have been produced in the same total number of collisions, but only a chance. It was clear that to discover the Z^0 the collider would have to reach a higher luminosity, which would also help to confirm the existence of the W particles by producing them more often.

Part of the problem of the collider's low luminosity is that the accumulator does not supply as many antiprotons as anticipated, storing at most some 1.2×10^{11} particles, one fifth of the number hoped for. This means that the best luminosity the physicists can expect is in the region of 2×10^{29} cm^{-2}s^{-1}. When the collider started up again on 12 April 1983, the aim was to improve its performance sufficiently to reveal the Z^0. By the end of May a new record in luminosity of 1.6×10^{29} cm^{-2}s^{-1} had been achieved, by using the low-beta magnets to squeeze the beams still tighter. But even earlier in the month the improved rate had been enough to provide UA1 with just what the physicists were looking for.

The Z^0 appears

On 4 May, out of the mass of data emerged one extremely interesting event, evidently containing a high-energy electron and positron, emerging back to back from the collision zone, just as expected for the decay of a Z^0 (Figure 58a). By the end of May, with more data and more candidates for Z^0 decays, the UA1 team had become more confident and they announced their discoveries to the press on 1 June. Indeed, in the period from 30 April to 28 May, UA1 had found no less than five likely Z^0s.

As with the W decays, behind the jubilant announcements of the Z^0s lay an immense amount of hard work necessary to pull the signals characteristic of the decays out of the great mass of information. The Z^0 is expected to decay to an electron–positron pair or a $\mu^+\mu^-$ pair. To this end, UA1 set up an analysis 'express line', with four computers working independently during data taking to put onto tape events with high transverse energy (more than 12 GeV) in the electromagnetic calorimeter, or

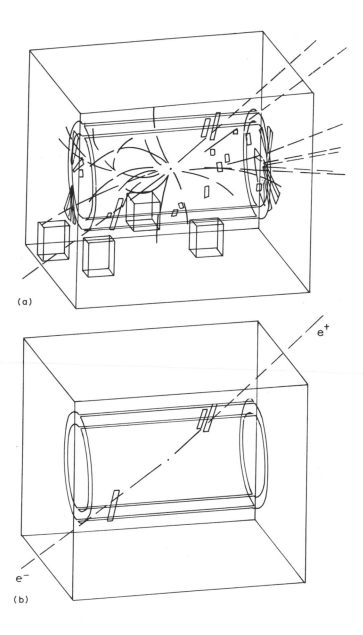

Figure 58 *A Z⁰ decays in the apparatus of UA1, having been created in a collision between proton and antiproton. The computer graphic display for the event contains several tracks (a), but when those with low energy are removed, only two remain (b), those from the electron and positron of the Z⁰ decay. On the energy distribution for the same event, the electron and the positron show up as two clear spikes (c). (From: G. Arnison et al., Physics Letters, vol. 126B (1983), p. 398)*

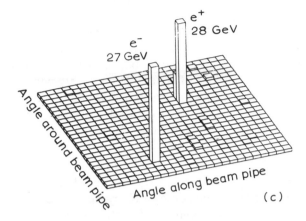

(c)

at least one wide-angle penetrating muon track from the interaction zone. From these events, the analysis chain searching for Z^0s selects only those with at least two clusters with a total transverse energy of more than 25 GeV deposited in cells in the electromagnetic calorimeter, and with little energy (less than 800 MeV) in the adjacent section of the hadron calorimeter. These criteria picked out only 152 events from the original sample.

The next step is to check that there is an isolated track at least 40 cm long, corresponding to a momentum of more than 7 GeV, in the central drift chamber and that this points to one of the clusters. At this point in the analysis only six events remained; two of these were rejected as they failed to satisfy the demand that the second cluster of cells in the electromagnetic calorimeter had a similar isolated track pointing to it. The four remaining events all seemed to show no signs of missing momentum, as the total measured transverse energy balanced; the two clusters of energy deposited in the electromagnetic calorimeter show up well as two spikes in plots of energy deposition (Figure 58c).

Those events recorded with penetrating muon tracks are analysed to check whether at least one track can be reconstructed in the muon chambers, tying up with a track in the central detector at least 40 cm long, with a measured transverse momentum of more than 7 GeV. These demands left forty-two events, which were then carefully studied individually to search for those events with two isolated muon tracks. Only one event finally passed this scrutiny; the other events could be explained as due to cosmic-ray muons.

From the measured energies and momenta of the electrons (or muons), the UA1 team can derive an estimate of the mass of the Z^0 particle, this time without any problems due to missing energy, so there is no need to fall back on comparison with theory. The five events give an average value of 95.2 ± 2.5 GeV, which, taken together with the mass UA1 found for the W particle, is in line with the electroweak theory.

UA1 confidently announced discovery of the Z^0 at the beginning of June, and immediately submitted a paper for publication in *Physics Letters*, based on the sample of data collected between 30 April and 28 May. The UA2 collaboration preferred to bide its time and did not make public its findings on the Z^0 until 15 July, by which time data taking on the collider had come to an end, and the team had found eight likely Z^0s.

UA2's analysis selects those events – out of a total of some 70,000 – that appear to contain electrons in the debris of the proton–antiproton collision; electrons are recognized by a characteristic deposition of energy in the electromagnetic calorimeter and in the 'preshower' counter before the calorimeter, together with a track in the central wire chambers. These events pass the next stage only if the transverse electromagnetic energy is above 30 GeV, and a pair of energy clusters corresponds to a mass greater than 50 GeV. From the initial data sample, 7427 events survived this far. These events are then checked to ensure that there is little hadronic energy deposited, and that the electromagnetic cluster is not too broad. Together these steps reduced the number of events to twenty-four.

Still tighter constraints to identify electrons reduced the total number of events to eight, all of which seem likely to contain the decays of a Z^0 (Figure 59). Four of the events completely satisfy all the criteria on both electrons of the Z^0 decay. Four others pass all criteria on only one of the two electrons. However, it is difficult to see what else they could be. The team therefore chooses to base its value for the mass of the Z^0 on the four remaining events, and finds an answer of 91.9 ± 1.3 ± 1.4 GeV, where the errors are in the measurements and in the precise energy calibration respectively.

On 3 July 1983, the beams of protons and antiprotons from the SPS were dumped and the second historic run came to an end. By this time UA1 had recorded a total of fifty-two clear W particle decays into eν and about twenty decays to $\mu\nu$, and eight Z^0 decays (four into e$^+$e$^-$ and four into $\mu^+\mu^-$). After further calibration work, the calculated masses turned out to be 80.9 ± 1.5 GeV for the W particle, and 95.6 ± 1.4 GeV for the Z^0. In both cases the errors quoted are statistical, and an additional uncertainty in the calibration of the energy scale of ±3 per cent applies. These values compare well with the electroweak theory (see the end of chapter 7), although the mass measured for the Z^0 is a little higher than predicted. The mass of the W particle can be used to calculate the parameter sin^2 θ_W, which is unpredicted by electroweak theory although

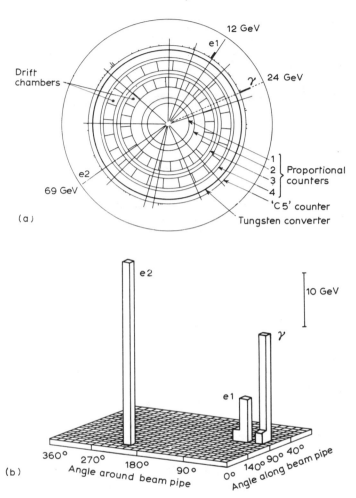

Figure 59 *The vertex detector of UA2 records the decay of a Z⁰, in which a photon is radiated, carrying away a substantial fraction of the total energy. An 'end on' view of the vertex detector (a) shows the tracks of the electrons (e1 and e2) and the photon (γ), all of which produce sharp peaks in the distribution of energy with angle (b). (From P. Bagnaia et al., Physics Letters, vol. 129B (1983), p. 130)*

it had been derived from many kinds of measurement, as chapter 7 describes. Using UA1's value for the mass of the W gives $\sin^2 \theta_W = 0.226 \pm 0.008$, with a systematic error of ± 0.014; this agrees well with the value of 0.233 ± 0.009 from other kinds of data.

Lastly, it is possible to calculate another parameter, called ρ, which relates the mass of the W and the mass of the Z^0 to the Weinberg angle, θ_W. This parameter should be equal to 1 in the simplest version of the electroweak theory in which only one scalar (spin-0) Higgs particle is required in the mechanism that breaks or 'hides' the symmetry between the electromagnetic and weak forces. More complex theories involve more scalar particles, but UA1's results give $\rho = 0.925 \pm 0.05$, suggesting that the simplest version of the electroweak theory will do perfectly well.

By 3 July 1983, UA2 had increased its number of W particles to thirty-five, giving a mass of 81 ± 2.3 GeV. This corresponds to $\sin^2 \theta_w = 0.226 \pm 0.014$, which can then be combined with the value measured for the mass of the Z^0 given earlier to give $\rho = 1.004 \pm 0.052$, again perfectly in line with the demands of the electroweak theory. The summer of 1983 had become the summer of intermediate vector bosons; the W and Z^0 could now move from the list of long-sought hypothetical particles to the list of real particles, with masses and other properties measured experimentally.

Other exotica

The discoveries of W and Z^0 particles had been the main aim in building the collider at CERN, but there were other perhaps more exotic possibilities that the experimental teams had kept in mind when designing the apparatus to study the collisions. As chapter 9 describes, there were to be searches for single quarks, freed from the confines of a hadron; for monopoles – particles carrying a single magnetic pole; and for the strange Centauro events that had been observed in the interactions of high-energy cosmic rays. What became of these investigations while the whole world seemed to celebrate the discoveries of the weak bosons?

Unfortunately, at least in these first few months of data collection at the proton–antiproton collider, nothing spectacular turned up. The four physicists in the UA3 team left their Kapton foils wrapped up around the heart of UA1 for the whole of the first period of collisions from August to December 1981. After immersing the foils in acid for four hours to etch away the tracks of ionizing particles, and searching for deep, broad holes, the researchers were forced to conclude that they had found no hole 'which could be attributed to the passage of a highly ionizing and penetrating particle'. In other words, there were no signs of a monopole.

The UA5 team was in search of other novel effects. These researchers installed their streamer chambers in the location later occupied by UA2 from October to November 1981, and during this time they recorded some 16,000 collisions. These data provided a first glimpse of the general properties of collisions in the new energy region that the collider had opened up, but of particular interest was the search for the so-called

Centauros – interactions that produce high numbers of charged particles, but hardly any neutrals.

To look for possible Centauros, the team analysed a selection of 1860 events, and found the average number of charged particles produced and the average number of photons. A photon would reveal itself by converting to an electron–positron pair, and initiating an electromagnetic shower, either in the stainless-steel wall of the beam pipe, or in lead-glass plates in one of the streamer chambers. Calculations based on the original Centauro events, found in emulsions exposed at high altitudes to cosmic rays, suggested that the chambers of UA5 might pick up around thirty to forty charged particles in an equivalent angular range. But a sample of events chosen particularly for a high multiplicity of charged particles (more than fifty) showed nowhere near as many as thirty to forty at angles corresponding to those covered by the cosmic-ray experiment. The team came to the conclusion that 'no high multiplicity events having features corresponding to Centauro events . . . have been observed'.

Another search for Centauros has been made by the UA1 team, whose all-purpose detector is certainly suitable for picking out such unusual events. The analysis in this case studied the energy deposited in the electromagnetic and hadron calorimeters, for a Centauro event should be characterized by a large amount of hadronic energy (mainly from charged pions) and little electromagnetic energy (from photons produced in the decays mainly of neutral pions). The information on the number of tracks in the central image chamber and the corresponding momentum was also valuable. The team had to interpret what the cosmic-ray events would look like in the UA1 detector, but after detailed considerations they, too, found no events with Centauro-like features in the sample of 48,000 that they selected to analyse.

Another breed of particle often searched for is the quark, the basic building block of all hadrons. While it is certainly true that quantum chromodynamics, the field theory of quarks and the strong nuclear force, implies that quarks can never appear as single entities but must always be found together in hadrons, there are some versions of the theory that suggest that single quarks might be freed at sufficiently high energies. It therefore seemed reasonable to check collisions at the proton–antiproton collider to discover if its new energies were high enough. To this end, the UA2 team made use of the fact that a segment of its main calorimeter can be removed to leave open an angle of 60° to one side of the collision zone, where other apparatus can be inserted.

In the search for quarks the main piece of equipment was a 'telescope' of scintillation counters, installed behind some wire chambers and an iron plate. The aim was to detect quarks through their low ionization compared with normal hadrons: the ionizing power of a charged particle is proportional to the square of its charge, so quarks with either $1/3$ or $2/3$ the charge of an electron produce only $1/9$ or $4/9$ the ionization. The analysis programs this time scoured the data collected for particles traversing the telescope, searching for those particles of abnormally low ionization. A few (twenty-three) events which fitted this criterion were found, but when these were examined carefully, in each case some explanation could be found for the nature of the event without having to invoke the existence of a free quark. Once again, there was no evidence in UA2 for particles with charge a fraction of that of the electron; it seems the quarks remain firmly in their bunches of twos and threes even at the newfound high energies of the collider.

This chapter has concentrated mainly on the search for the W and Z^0 particles at the proton–antiproton collider – after all, these are the particles the device was essentially *supposed* to find, such was confidence in the electroweak theory. The search for other previously unobserved particles has been less fruitful, but this should not obscure the fact that the collider has played an important role in extending our view of what we can call 'conventional physics', to distinguish it from the more unconventional studies. Here the machine entered an energy region previously explored only with cosmic rays, and it has already provided useful checks on theories of the strong force and the fundamental nature of particles. The collider's role is far from over,* at least until a machine is built that supersedes its energy. There is plenty still for the teams of experimenters to study, but for the theorists the path lies in looking beyond electroweak theory, a path down which many have already been exploring for some years.

* It is due to start work again in the autumn of 1984.

11

On to a Better Theory

The successes of CERN's proton–antiproton collider have in a sense closed one chapter in the history of particle physics. The demonstration that intermediate vector bosons can be created and observed to decay as real particles has brought to maturity ideas concerning the weak nuclear force that were founded in the 1930s, with the work of Yukawa, Klein and Fermi. Moreover, the discovery of these particles has strengthened the belief that gauge-field theories provide the correct description of nature's fundamental forces and the particles of matter they bind together. But amidst all the excitement of the past few chapters, it should not be forgotten that the theoretical ideas I have presented fall quite a long way short of providing a complete description of the basic physics of our universe. Fortunately, theorists need not be fettered by lack of experimental evidence and for several years many have been studying ideas that go far beyond the electroweak theory – ideas that tie together not only the electromagnetic and weak forces, but also the strong force and, in the most advanced theories, the gravitational force too.

This theoretical confidence, which seems sometimes to border on conceit, should not hide the fact that even the electroweak theory contains some loose ends. Experiments have yet to reveal the existence of the Higgs particle. This scalar (spin-0) particle is required as witness to the symmetry-breaking mechanism which means that in the real world we observe weak and electromagnetic effects as manifestly different facets of the same underlying electroweak force (see chapter 6). The Higgs mechanism is crucial to the theory, giving the W and Z^0 particles their masses in a way that keeps the theory free from intolerable infinities and hence ensures the property of gauge symmetry. Without evidence for the Higgs particle, the theory still stands on somewhat shaky ground. With the evidence for the Higgs, there will remain the question of just which version of the basic electroweak theory is correct; are there more than one Higgs particle, as some versions require? CERN's collider may solve these problems soon, or in a few years' time Fermilab, in the USA, may provide the answers when it produces proton–antiproton collisions at still higher energies in its new 1000-GeV superconducting magnet ring.

Another fundamental particle awaiting proof of its existence is the top quark. Recall from chapter 3 that all the hadrons observed so far appear to be built from a selection of quarks of five different kinds, or 'flavours'. Flavour quantum numbers, such as strangeness (s) and charm (c), characterize the different quarks. And in chapter 6 we saw how Glashow, Iliopoulos and Maiani (GIM) found that they could describe correctly the weak interactions of quarks of three flavours ('up', 'down' and 'strange') only if a fourth flavour, 'charm', existed. Subsequently (chapter 7) hadrons containing charmed quarks were discovered, but only a few years later, in 1977, evidence appeared for yet another quark flavour, this time dubbed 'bottom', or 'beauty' by the more poetic physicists. The original work of Glashow *et al.*, had not envisaged still more kinds of quark – suggesting a fourth variety was a bold enough move in itself. But the bottom quark does not provide too many problems, providing nature's blueprint includes provisions for a sixth flavour, 'top' or 'truth'. Then the GIM mechanism for calculating the weak decays from quarks of one flavour to lighter quarks of a different flavour can be successfully extended to operate between six flavours.

The standard picture of 'quantum flavourdynamics' describes the weak interactions of the flavoured quarks in terms of the formalism of the electroweak theory. The quarks and the leptons fall into doublets of the same 'weak isopin' – a property introduced in chapter 6. In each lepton doublet, a charged lepton is coupled with its respective neutrino – the electron resides with the electron–neutrino, the muon with the muon–neutrino and the tau with its (as yet undiscovered) partner, the tau-neutrino. As for the quarks, the up quark pairs with the down quark, the charmed quark with the strange quark, and the undetected top quark with the bottom quark. Strictly speaking, the picture is not quite so simple, for the weak force operates between mixtures of quarks, and some members of the doublets are states of mixed flavour, rather than a single flavour. However, the basic idea to appreciate is that the standard picture of flavourdynamics looks good if there are six flavours rather than five, so the top quark should – some would say must – exist.

CERN's collider may again provide the desired effects, but searching for the telltale decays of hadrons containing top quarks, among the millions of events containing many hadrons, is more difficult than searching for the relatively clean signature of the demise of a W or Z^0 particle. As with the other five flavours, the top quark will not be seen directly in the experiments, for it seems that quarks cannot exist in isolated states. Rather, the proton–antiproton collisions at CERN should have sufficient energy to create quark–antiquark pairs endowed with the labels – quantum numbers

– of top and antitop. These top quarks and antiquarks will materialize as hadrons just as the more familiar quarks do. A 'top particle', that is, a particle carrying the flavour quantum number called top – will be a hadron containing a top quark, say, together with another type of antiquark (to form a meson) or two other quarks (to form a baryon)*. The top quark must be heavier still than the bottom quark, which gives rise to the heaviest hadrons discovered so far, with masses some ten times that of the proton. Such a particle should reveal itself through a cascade of weak decays, where the flavour quantum number changes at each stage, ultimately leading to configurations of up and down quarks, that is, to pions, protons and neutrons. Once the top quark is found, its properties, such as its mass and lifetime, will provide a crucial test of the way that electroweak theory is applied to quarks.

Problems with the standard model

The existence of a quark of yet another flavour may tidy up some areas of theory, but it begs a rather more fundamental question that theory at this level fails to address: why are there so many flavours? And have we found all the flavours there are? All we need to build the materials of the everyday world, including the building bricks of our own bodies, are the 'up' and 'down' quarks – these are all that are necessary to form the protons and neutrons in atomic nuclei. Similarly, the only leptons we need are the electron, together with its neutrino, which we need in the weak interactions that fuel our sun. The muon, and its neutrino, and the strange quark appear only when we begin to study the high-energy reactions of cosmic rays in the upper atmosphere; they play no role in building the world about us. Likewise, the charmed and bottom quarks, and the tau lepton have appeared only in our explorations at particle accelerators, explorations that have sought to elucidate the true nature of matter. Yet these unasked for 'elements' seem to play an important role in nature's scheme; the problem remains to understand just what this role is.

There are other unsatisfactory points about what is generally termed the 'standard model', comprising electroweak theory together with quantum chromodynamics (QCD), the best theory we have for the strong nuclear force (see chapter 4). Why, for example, is electric charge quantized, with the proton's charge the same size (but opposite sign) as the electron's? This

---

* In July 1984, there emerged the first evidence for the decay of a top quark, observed in the UA1 detector at CERN's proton-antiproton collider. In the few events in question, the detector reveals an energetic lepton (muon or electron), two jets of hadrons, and missing energy consistent with the production of a neutrino. This pattern is precisely what one expects if a W particle, formed in the proton-antiproton collision, decays into a top quark (or antiquark) together with, say, a bottom antiquark (or quark).

comes down to asking why the quarks have charges of $2/3$ and $1/3$, and leptons have charges 0 and 1 in units of e, the charge of an electron. Electroweak theory does not say what these charges should be; they have in effect to be inserted 'by hand'. Moreover, the masses of all the quarks and leptons are quite arbitrary, as are the strengths of the interactions; recall that electroweak theory makes no prediction for the value of $\sin^2\theta_W$, the parameter that relates the strengths of the electromagnetic and weak forces. These and other arbitrary parameters make the standard model generally unsatisfying. After all, a complete theory should surely say why the universe is the way it is, rather than need to have numbers describing the universe inserted into it.

Before I proceed to indicate the directions in which theorists are working in order to solve these problems, I should recap on the 'standard model' of electroweak theory and QCD. Electroweak theory (a gauge theory) provides, as we have seen, an excellent description of the weak and electromagnetic forces; QCD (also a gauge theory) provides a pretty good theory of the strong interactions. The story of QCD belongs to another book, but I can summarize briefly the theory's main points, which were introduced in chapter 4. The strong force, according to QCD, is manifest through a field arising from a property unique to quarks. This property has the rather confusing name of colour, because it seems to have three 'values', like the three primary colours of light. The important point to remember is that colour is to the strong force as electric charge is to the electromagnetic force – it is the property of the particles that is the source of the force and its field. Particles without colour, such as leptons, do not feel the strong force, just as uncharged objects do not feel the electromagnetic force.

A second point is that the strong force is carried by vector bosons, like the W and Z^0 particles of the weak force and the photon of electromagnetism. The vector bosons of the strong field are called gluons, and they come in eight varieties. The main difference between a gluon and a photon, say, is that while the photon is uncharged, the gluons carry colour. This means that while photons cannot be bound together by the electromagnetic force to form 'atoms' of light, gluons can mutually interact and combine to form composite structures, or 'glueballs'. This property of gluons is also intimately linked with the confining nature of the strong force – the fact that it becomes stronger at greater distances between particles, and therefore keeps quarks confined within the hadrons they form.

The gauge theory of the strong force, QCD, describes the behaviour of the coloured quarks and the coloured gluons. As far as QCD is concerned, the *flavour* of a quark is not important; the strong force operates between quarks of three different colours and it is this action that QCD describes. The mathematical group of symmetry transformations in the theory is the same group, SU(3), that Murray Gell-Mann and Yuval Ne'eman used to build particles from three constituent quarks. To distinguish the two uses, we can write SU(3)$_c$ to refer to the group describing three coloured quarks.

The grand unified solution

The standard model thus contains the groups SU(2)xU(1), in the electroweak theory, and SU(3)$_c$ in QCD; together they can be written as SU(2)xU(1)xSU(3)$_c$. Although the standard model seems unsatisfactory, its successes, including the prediction of the W and Z^0 particles, are sufficient to suggest that it describes at least some aspects of the observable world with reasonable accuracy. A number of theorists have thus been drawn to the idea that a better theory is a field theory that contains this standard model. Taking a lead from electroweak theory, which shows that the weak and electromagnetic forces are but two facets of the same basic force, it seems reasonable to construct a 'larger' theory in which the strong force is also a facet of the same underlying force. In other words, if the weak force and electromagnetic force are in some sense symmetric – it is just that their symmetry is hidden from us in our everyday world – then it seems reasonable that the strong force is also part of a grander symmetry.

There have been several attempts at 'grand unified' theories that seek to link the three forces, apart from gravity. The one most discussed, in part because it is conceptually the simplest, is due to Sheldon Glashow and his colleague at Harvard, Howard Georgi. It is some indication of how theorists can forge ahead unconstrained by want of experimental evidence that Georgi and Glashow first published this work in 1974, when the experimental view of the particle zoo was quite different from today. The mathematical group that Georgi and Glashow suggested to contain the fundamental quarks and leptons is called SU(5), and its basic representations can be identified with the quarks and leptons of one 'generation' – for example, the up and down quarks (each in three colours), their antiquarks, the electron, the positron and the electron–neutrino. The SU(5) symmetry group then gives all the transformations possible between the different elements of the group, that is, between the quarks and the leptons it contains.

It turns out that the SU(5) model provides answers to some of the questions that the standard model leaves wide open. It fixes the charges of the quarks and the leptons at their different values; it relates the fractional charge of the quark to the number of colours; and it provides relationships among the different coupling strengths of the three observed forces. This all seems quite remarkable, so what price do we have to pay?

Electroweak theory required the existence of four gauge bosons: the photon, the two charged W particles and the Z^0 particle. SU(5) requires twenty-four gauge bosons. Four of these can be identified with the four of electroweak theory, and a further eight correspond to the eight gluons of QCD. That leaves twelve 'new' vector bosons, which mediate processes unsuspected in the standard model, for SU(5) brings quarks and leptons together as equal members of the same basic multiplets. This means that quarks should be able to transform not only into other quarks but also into leptons – a proposition that strikes at the very heart of our notions of the stability of matter.

The additional vector bosons of SU(5) mediate quark–lepton transmutations that violate the empirical laws of baryon and lepton conservation introduced in chapter 3. Experiments suggest that baryons and leptons are always created or destroyed in pairs with their antiparticles, so that the total baryon number, and the total lepton number, remains the same. But if a proton can decay to a positron, as SU(5) suggests, then these laws no longer hold.

What evidence is there that the proton can decay? At the time of writing, there is essentially none. We know, as Maurice Goldhaber of the Brookhaven National Laboratory has pointed out, that the lifetime of a proton must be at least 10^{16} years, otherwise our bodies, each containing some 5×10^{27} nucleons, would be radioactive sources, emitting more radiation than the authorities allow for radiation workers! However, SU(5) predicts proton lifetimes much longer than this.

The reason for the proton's relative longevity is related to the gauge bosons that mediate its decay, and to the strengths of the strong, weak and electromagnetic interactions as we observe them. One consequence of the field theories of these interactions is that the strengths vary with the energy of the interaction. In particular, the strong interaction becomes weaker at higher energies, and one can envisage some high energy at which the three interactions become of equal value. This indeed echoes some of the philosophy of the electroweak theory. In the low-energy world we inhabit, weak and electromagnetic effects appear quite different, due to the large difference between the massless photon and the massive W and Z^0 particles. At high energies, like those of CERN's proton–antiproton collider, W and Z^0 particles can materialize and begin to play a role equally important to that of the photon: the underlying symmetry is no longer so obscure. In a similar way, SU(5) predicts that at even higher energies the symmetry between electroweak and strong effects will be unveiled, and the additional bosons will begin to play their roles.

The strengths of the interactions seem to converge at an energy of around 10^{15} GeV, some 10^{13} times the energy needed to make the W and Z^0 bosons materialize, and this gives a clue to the nature of the new bosons. They must be extremely heavy indeed, some 10^{15} times the mass of the proton, and even at the energies of the proton–antiproton collider (540 GeV) we have no chance of observing their effects. Moreover, the great mass of the superheavy bosons means that proton decay is indeed a very rare process and, depending on the fine details of the theory, one can calculate a lifetime of some 10^{32} years.

Such large numbers do not deter experimenters, who have set out to build detectors containing sufficiently large numbers of protons – 1000 tonnes of water contain around 10^{33} protons – that they may observe a few decays per year. There are now a number of detectors in operation, in the USA, Europe, Japan and India. Already, occasional events have been observed that could indicate the decay of a proton, but there is insufficient evidence to say conclusively that protons have been found to decay. Indeed, one measurement indicates that the lifetime for the decay to a positron and a neutral pion must be at least 6.5×10^{31} years.*

SU(5) is by no means the only grand unified theory, but it serves to illustrate the basic features of such theories. The intimate linking of quarks and leptons almost invariably leads to predictions of proton decay. Moreover, it is clear that the grand unification must occur at very high energies, unfortunately well beyond the highest energy that accelerators are likely to reach, which makes experiments such as those that search for proton decay all the more important. However, grand unification is by no means the end of the line for the theorists.

The supersymmetric solution

Like the standard model before them, the grand unified theories, even if they do prove correct in predictions of proton decay, for example, still leave some questions unanswered. Their 'unification' generally refers to those particles, the quarks and leptons, of one generation: there is still no answer to the questions of why there is more than one generation, or how many generations there are. Furthermore, the theories still contain an undesirable number of arbitrary constants, such as masses and the so-called mixing angles that determine the rates of reactions. In addition, the grand unified theories create a puzzle of their own, namely, why is there such a large gap between the energy scale for the first step in unification between weak and electromagnetic forces (around 100 GeV) and the second step in which the strong force joins in (around 10^{15} GeV) – a gap of 13 orders of magnitude, or powers of ten?

The solution to these problems may turn out to lie in a class of theories that began to emerge in the early 1970s. Here again we can see how the theorists must be continually chipping away at new territory, directed sometimes only vaguely by experiment, if progress in the long term is to

---

* R. M. Bionta *et al.*, *Physical Review Letters*, vol. 51 (1983), p. 27.

be achieved. These theories, which have lately become very fashionable, are called supersymmetric theories and they have a number of features that make them particularly attractive. Their main property is that they link *all* kinds of particle – the fermions of half-integral spin (the quarks and the leptons), the spin-0 objects such as the Higgs particles, and the vector bosons with spin 1. Thus the supersymmetric theories seem to bridge the present division between the spin-$\frac{1}{2}$ (matter) particles and the spin-1 bosons that mediate the forces between the matter particles.

Once more, the elegance of the theories comes at a price that must be tested against observations in the real world. The grouping together of bosons and fermions means that the one can transform into the other; this in turn implies that for all the known fermions (the quarks and the leptons) there must exist supersymmetric partners which are bosons, in fact spin-0 scalars. These have been dubbed 'squarks' and 'sleptons', where the initial letter *s* indicates a scalar particle. Similarly, there should be fermionic counterparts to the known bosons: particles with spin $\frac{1}{2}$, which have been named the photino, zino, wino and gluino, in obvious fashion. If the supersymmetry were exact, then the masses of the super-symmetric partners would be exactly equal, but experiments show that this at least cannot be right. For example, we know that there are no scalar particles with the mass of the electron, or the muon. The symmetry of supersymmetry must be broken, or hidden in our low-energy world, though this, of course, poses no insuperable problems for theorists.

But there are some more important reasons why supersymmetry has become so exciting. Perhaps the most significant of these is that it is possible to form a supersymmetric theory that includes gravity, the weakest of all the forces. Gravity, the force that is easiest to observe through its influence on apples, cricket balls, and so on, and which was the first force to succumb to a reasonable theoretical treatment with the work of Newton, proves to be the most difficult of the four forces to fit into the ideas of quantum-field theory – ideas that seem to have so much success with the other three forces. It is not that gravity cannot yield to a clean theoretical treatment. Einstein's theory of general relativity provided a gauge theory of gravity long before the advent of the elec-troweak theory. The problem with gravity comes when one tries to mesh general relativity with quantum theory, and the matter fields of quarks and leptons. Such theories are impossible to renormalize, with numerous infinities at every turn in the calculations.

Unfortunately, the situation does not improve with the simplest form of supergravity, the supersymmetric theory of gravity. Such a theory includes a graviton of spin 2 – the boson supposed to mediate the gravita-tional force – and its supersymmetric partner, the gravitino, which should have a spin of $\frac{3}{2}$. There does, however, seem to be some important relation between supersymmetry and gravity, for the version of supersymmetry that is *locally* symmetric, that is symmetric under different transformations at different points in space and time, generates the very spin-2 particles (gravitons) that a quantum theory of gravity requires. The problems of the infinities can, however, be overcome in a more complex version of supergravity, known as extended supergravity. There are different versions of this depending on the basic number of supersymmetric charges, N, but the most popular with theorists has $N = 8$. This is the largest value N can take without the theory giving rise to particles with spin greater than 2, which nature seems not to require.

In '$N = 8$ extended supergravity' there is one spin-2 particle (the graviton), eight with spin $3/2$ (all gravitinos), twenty-eight spin-1 particles, fifty-six with spin $1/2$ and seventy scalars with spin 0. Surprising as it may seem, this abundance of particles does not seem sufficient to identify all the known bosons and fermions, in particular the W^+ and W^- bosons! With $N = 8$ extended supergravity we have not yet found the right answer. Or have we?

One of the most tantalizing features of the theory is that it seems far less troubled with infinities than other theories that attempt to incorporate gravity at the quantum level. The effects of the 'new' particles such as the gravitinos tend to cancel out infinities that arise from possible interactions between the more conventional particles. Whether *all* the infinities will cancel out is not yet known, but the theory offers the hope that this may be the case. If the infinities do cancel out, it will be a tremendous achievement. It will mean that the theory is rid of infinities without recourse to a renormalization procedure like that used in QED, which manages in some sense to 'hide' the infinities in the mass and charge of the electron, parameters defined by measurement. A theory in which the infinities cancel naturally would be entirely self-consistent and would show that nature is wholly determined by its underlying symmetry.

One way round the problem of fitting all the quarks and leptons into the theory may be that we have not yet found the fundamental constituents of matter; that there exist a number of 'prequarks' for instance, which *can* be identified successfully within the supermultiplet of $N = 8$ extended supergravity. In many respects the theorists have the freedom to toy with many ideas, unlimited by the scope of the apparatus, or the energies that they can probe. The universe is their laboratory, from the big bang of creation, to a future in which all protons have perhaps decayed to leptons, and quarks no longer exist. The conditions we observe in the universe at present, out to its farthest reaches, and the effects we find in experiments, are the raw material the theorists have from which to mould their unified model of the physical world. The W and Z^0 particles are important structural members of that model. The challenge is to seek out more.

Appendix I
Numbers – Big and Small

Throughout this book you will come across examples of the exponential or 'scientific' notation for writing down numbers that are either very large or very small. Where possible, I have tried to avoid this, and I have spelt out 'ten thousand million', or 'one hundred-millionth', for example. (I have also purposely not used the word billion, to avoid confusion on either side of the Atlantic.) However, some numbers in particle physics are either so small, or so large, that to express them in words becomes nigh on impossible. In such cases I have used the standard scientific shorthand of exponential notation.

Numbers larger than 1 can be written in terms of powers of ten: $10 = 10^1$, $100 = 10^2$, $1000 = 10^3$, and so on. Thus 10^{32} is the figure 1 followed by 32 noughts. Using this notation I can therefore express the lifetime of the proton (chapter 11, p. 167) as 'at least 6.5×10^{31} years', where the total number of years is 6 followed by 5 and then 30 noughts (the 5 takes the place of a nought).

In a similar fashion, numbers smaller than 1 can be expressed as negative powers of ten: $0.1 = 10^{-1}$, $0.01 = 10^{-2}$, $0.001 = 10^{-3}$, and so on. In this case 10^{-31} is shorthand for a decimal point followed by 30 noughts, and then a figure 1. Equivalently, I could have written $10^{-31} = 1/10^{31}$, or 1 divided by 1 followed by 31 noughts. In this way I can write the mass of the electron (chapter 1, p. 11), as 9.1×10^{-31} kg $= 9.1/10^{31}$, or 9.1 divided by 1 followed by 31 noughts.

Appendix II
Units of Energy

The mass of a proton is very small: 1.7×10^{-27} kg; in other words you need 10^{27} protons to weigh 1.7 kg. This means that the proton can have an energy that is very high as far as the proton is concerned, but very small in everyday terms. The energy equivalent to the mass of one proton is 1.5×10^{-10} joules – a very small amount of energy; but 'only' 10^{19} protons (17 micrograms) converted every second would fuel an average power station.

To deal with individual particles, physicists use a unit called the electronvolt. This is the energy an electron (or a proton or anything of unit charge) gains when it is accelerated across a voltage difference of 1 volt. Einstein's equation $E = mc^2$, allows us to equate energy E with mass m, in terms of the velocity of light c. So we can express masses in terms of energy units, which turns out to be very convenient.

The mass of the electron is then 510 thousand electronvolts, or kiloelectronvolts; the mass of a proton is 935 million electronvolts, or megaelectronvolts. For convenience the units are broken up into groups, 1000 at a time, as follows.

1 kiloelectronvolt (1 keV)   = 1000 eV
1 megaelectronvolt (1 MeV) = 1000 keV = 1,000,000 eV
1 gigaelectronvolt (1 GeV)   =   1000 = 1,000,000,000 eV
                                MeV
1 teraelectronvolt (1 TeV)   = 1000 GeV = 1,000,000,000,000 eV (or 10^{12} eV)

Where possible I have tried to avoid using energy units, and have related particle masses to the proton's mass. But in some instances, such as when talking about accelerators, this is not really helpful and I have used the standard units.

Appendix III
The Greek Alphabet

Many subatomic particles are known not by special names, such as 'proton' or 'electron', but by Greek letters that have been assigned to them. This table lists the pronunciation of the Greek alphabet for those readers not familiar with it.

A	α	alpha (a)	N	ν	nu (n)
B	β	beta (b)	Ξ	ξ	xi (x)
Γ	γ	gamma (g)	O	o	omicron (o)
Δ	δ	delta (d)	Π	π	pi (p)
E	ε	epsilon (e)	P	ρ	rho (r)
Z	ζ	zeta (z)	Σ	σ	(ς final or C c) sigma (s)
H	η	eta (ē)	T	τ	tau (t)
Θ	θ	theta (th)	Y	υ	upsilon (u)
I	ι	iota (i)	Φ	φ	phi (ph)
K	κ	kappa (k)	X	χ	chi (ch)
Λ	λ	lambda (l)	Ψ	ψ	psi (ps)
M	μ	mu (m)	W	ω	omega (ō) or Ω

Further Reading

General background on particle physics

The Nature of Matter, edited by J. H. Mulvey (Oxford University Press, 1981). This is a series of lectures given at Wolfson College, Oxford, in 1980, by a number of well-known experts. Aimed at a general audience, the talks assumed little background knowledge of physics.

The Cosmic Onion: Quarks and the Nature of the Universe, by Frank Close (Heinemann, 1983). An account of particle physics in the twentieth century for the general reader.

Particle accelerators and scientific culture, by Ugo Amaldi (CERN, 79-06). An interesting account of the growth of particle physics by one of CERN's senior scientists, this includes many key references to original work and quotes from important figures.

Introduction to High Energy Physics (second edition), by Donald Perkins (Addison-Wesley, 1982). A more advanced textbook for the physics undergraduate or postgraduate.

'The Birth of Nuclear Physics' by Peter Robinson, *New Scientist*, vol. 93 (1982), p. 426. An account of the seminal discoveries of 1932.

'Particles Play the Generation Game' by Frank Close, *New Scientist*, vol. 84 (1979), p. 701. A short introduction to quarks, leptons and so on.

Electroweak theory

Texts of the speeches made by the recipients of the 1979 Nobel prize in physics give personal accounts of the development of electroweak theory. These can be found in:

'Gauge Theories of the Fundamental Particles' by Abdus Salam, *Science*, vol. 210 (1980), p. 723.

'Conceptual Foundations of the Unified Theory of Weak and Electromagnetic Interactions' by Steven Weinberg, *Science*, vol. 210 (1980), p. 1212.

'Towards a Unified Theory: Threads in a Tapestry' by Sheldon Glashow, *Science*, vol. 210 (1980), p. 1319.

Other interesting articles appear in:

'The 1979 Nobel Prize in Physics' by Sidney Coleman, *Science*, vol. 206 (1970), p. 1290. This gives some background to the development of electroweak theory.

'Intermediate Bosons: Weak Interaction Couriers' by P. Q. Hung and C. Quigg, *Science*, vol. 210 (1980), p. 1205. An account of the history of the W and Z^0 particles.

'Gauge Theories of the Forces between Elementary Particles' by Gerard t'Hooft, *Scientific American*, June 1980, p. 90. A good introduction to the ideas of gauge theories and their importance in physics.

'The Renaissance of Gauge Theory' by K. Moriyasu, *Contemporary Physics*, vol. 23 (1982), p. 553. A more advanced account of the use of gauge theories.

'After the W particle, the Higgs' by Andrew Watson, *New Scientist*, vol. 98 (1983), p. 534. A clear account of the Higgs mechanism of spontaneous symmetry breaking.

The antiproton collider

'The Search for Intermediate Vector Bosons' by David Cline, Carlo Rubbia and Simon van der Meer, *Scientific American* (March 1982), p. 38. Three of the physicists instrumental in the collider project describe its development.

'Antiproton-Proton Colliders and the Intermediate Vector Bosons' by David Cline and Carlo Rubbia, *Physics Today* (August 1980), p. 44. A more technical review of the workings of the collider and mechanisms for cooling antiprotons.

The discovery of the W and Z^0 particles

'The Quest for the W Particle', *New Scientist*, vol. 97 (1983), p. 221, and 'CERN Physicists Find the Z^0 Particle', *New Scientist*, vol. 98 (1983), p. 355, by Christine Sutton. Two articles written immediately after the first announcements of the discoveries at CERN.

'Found: W and Z' by Alan Astbury, *Physics Bulletin*, vol. 34 (1983), p. 434. An account by one of the leading members of the UA1 team.

The details of the discoveries by the UA1 and UA2 teams are published in the following papers:

G. Arnison *et al.*, *Physics Letters*, vol. 122B (1983), p. 103, and G. Arnison *et al.*, *Physics Letters*, vol. 129B (1983), p. 273. These give details of W decays in UA1.

M. Banner *et al.*, *Physics Letters*, vol. 122B (1983), p. 476, gives details of W decays in UA2.

G. Arnison *et al.*, *Physics Letters*, vol. 126B (1983), p. 398, describes Z^0 decays in UA1.

P. Bagnaia *et al.*, *Physics Letters*, vol. 129B (1983), p. 130, describes Z^0 decays in UA2.

Beyond the electroweak theory

The ideas mentioned briefly in chapter 11 are continually evolving. Here are some articles that show the direction in which theorists are heading.

'The Structure of Quarks and Leptons' by Haim Harari, *Scientific American* (April 1983), p. 48.

'Waiting for the End' by Christine Sutton, *New Scientist*, vol. 85 (1980), p. 1016.

'Grand Unification: Tomorrow's Physics' by Sheldon Glashow, *New Scientist*, vol. 87 (1980), p. 869.

'Unified Theory of Elementary-Particle Forces' by Howard Georgi and Sheldon Glashow, *Physics Today* (September 1980), p. 30.

'A Unified Theory of Elementary Particles and Forces' by Howard Georgi, *Scientific American* (April 1981), p. 40.

'Is the End in Sight for Theoretical Physics?' by Stephen Hawking (Cambridge University Press, 1980). The author's inaugural lecture as Lucasian Professor of Mathematics at Cambridge University, it provides a farsighted look at the future in store for theories of fundamental forces and particles.

'The Decay of the Proton' by Steven Weinberg, *Scientific American* (June 1981), p. 52.

'Universal Supersymmetry' by James Dodd, *New Scientist*, vol. 83 (1979), p. 597.

'Supersymmetry and Supergravity' by Mitchell Waldrop, *Science*, vol. 220 (1983), p. 491.

Acknowledgements

The biggest 'thank you' must go to all the physicists and engineers at CERN who made possible the project that lies at the heart of this book. I must single out especially all those members of the UA1 team with whom I had the opportunity to work all too briefly in the early days which now seem so long ago. Especially, I must mention Mirella Keller, who keeps the UA1 machine operating smoothly, and who has suffered innumerable phone calls from me, begging her assistance in tracking down one wayward physicist or another. Then I would like to thank all those who have checked through chapters for me, and who can all claim their free beers now. Ian Aitchison, from Oxford University, deserves two beers for his detailed comments and for helping to fix the chinks (gaping holes?) in my theoretical armour. Others who have helped me cover my ignorance are Roy Billinge (CERN), Allan Clarke (CERN), Frank Close (Rutherford Appleton Laboratory), Paul Davies (Newcastle University), Jim Homer (Birmingham University), Peter Kalmus (Queen Mary College, London), Don Perkins (Oxford University, and whose book *Introduction to High-Energy Physics* provided invaluable guidance) and Andrew Watson (Gulf Polytechnic, and who is probably still suffering nightmares from the day he found my baby in his office). I must also thank Colin Tudge from *New Scientist* for helpful comments. Finally, many thanks to Mandy Caplin for her immaculate typing of my manuscript – she is far friendlier than any word processor I know – and to Oliver Caldecott who suggested I write the book in the first place. All those I have forgotten, please forgive me; if I ever do write another book, I'll squeeze you in next time round.